Foundations of Engineering Geology

Foundations of Engineering Geology

TONY WALTHAM
BSc, DIC, PhD

Civil Engineering Department
Nottingham Trent University

Second Edition

Spon Press

London and New York

First published 1994 by E & FN Spon

This edition first published 2002
by Spon Press
11 New Fetter Lane, London EC4P 4EE

Simultaneously published in the USA and Canada
by Spon Press
29 West 35th Street, New York, NY 10001

Reprinted 2002, 2003, 2004

Spon Press is an imprint of the Taylor & Francis Group

Typeset by RefineCatch Limited, Bungay, Suffolk
Printed and bound in Great Britain by
The Alden Group, Oxford and Northampton

British Library Cataloguing in Publiction Data
A catalogue record for this book is available from the British Library

Library of Congress Cataloging in Publication Data
A catalog record for this book has been requested

ISBN 0–415–25449–3 (Hbk)
ISBN 0–415–25450–7 (Pbk)

Preface

Civil engineering is an exciting combination of science, art, professional skill and engineering achievement which always has to rely on the ground on which its structures stand. Geology is therefore vital to success in civil engineering, and this book brings to the reader those many aspects of the geological sciences specifically relevant to the profession.

This book is structured primarily for the student of civil engineering who starts with no knowledge of geology but is required to understand the ground conditions and geological processes which, both literally and metaphorically, are the foundations of his future professional activities. It also provides an accessible source of information for the practising civil engineer.

All the material is presented in individual double-page spreads. Each subject is covered by notes, diagrams, tables and case histories, all in bite-sized sections instead of being lost in a long continuous text. This style makes the information very accessible; the reader can dip in and find what he needs, and is also visually guided into relevant associated topics. There is even some intended repetition of small sections of material which are pertinent to more than one aspect within the interrelated framework of a geological understanding.

The contents of the book follow a basic university course in engineering geology. The free-standing sections and subsections permit infinite flexibility, so that any lecturer can use the book as his course text while tailoring his programme to his own personal style. The single section summarizing soil strength has been included for the benefit of geology students who do not take a comprehensive course in soil mechanics within a normal civil engineering syllabus.

The sectionalized layout makes the information very accessible, so that the practicing engineer will find the book to be a useful source when he requires a rapid insight or reminder as he encounters geological problems with difficult ground. Reference material has therefore been added to many sections, mainly in tabulated form, to provide a more complete data bank. The book has been produced only in the inexpensive soft-bound format in the hope that it will reach as large a market as possible.

The mass of data condensed into these pages has been drawn from an enormous variety of sources. The book is unashamedly a derived text, relying heavily on the worldwide records of engineering geology. Material has been accumulated over many years in a lecturing role. A few concepts and case histories do derive from the author's personal research; but for the dominant part, there is a debt of gratitude acknowledged to the innumerable geologists and civil engineers who have described and communicated their own experiences and research. All the figures have been newly drawn, and many are derived from a combination of disparate sources. All the photographs are by the author, except for the Meridian air photograph on page 39.

Due thanks are afforded to the Department of Civil and Structural Engineering at the Nottingham Trent University where the engineering and teaching experience was gained, to Neil Dixon for his assistance with the gentle art of soil mechanics, to the staff of Blackie in Glasgow who made the innovative style of the book possible, and to the many colleagues and friends without whom nothing is possible.

T.W.

Preface to the Second Edition

The second edition of this book has been carefully updated and improved with additional paragraphs while keeping to the format and structure that has proved so accessible and so popular.

The one new section is #37, *Understanding Ground Conditions*, which has been included in an attempt at persuading the engineer to stand back and take a broader view of the overall geology at a site. Though this may seem to lack relevance in assessing the smaller details of a single urban building site, it does have real benefits in assessing ground conditions and evaluating potential geohazards on larger construction projects. The concept of the big picture is always useful, and this is very much the modern approach to engineering geology. Keeping to the same theme of contemporary geology, a box on brownfield sites has been included in the new section.

This book was never intended to be a handbook with all the answers and all the procedures. It is aimed to introduce the critical aspects of geology to the student of engineering, though it does appear to act as a convenient reminder to the practising engineer. To enhance its role as a source book, a long list of further reading has been added to this edition. It cites the useful key texts in each subject area, and also the primary papers on case studies used within the text, in both cases without any need to include conventional references that can disrupt a text.

As in the first edition, there are no cross references to other pages in order to explain terms being used. The index is intentionally comprehensive, so that it can be used as a glossary. Each technical term in the text does appear in the index, so that the reader can check for a definition, usually at the first citation of a term.

Sincere thanks are recorded to Peter Fookes, Ian Jefferson, Mike Rosenbaum, Jerry Giles and various others who have contributed to the revisions within this second edition, and also to the students of Nottingham Trent University who have road-tested the book and made the author appreciate the minor omissions and irritations that could be smoothed out.

T.W.

Contents

01 Geology and Civil Engineering

THE GEOLOGICAL ENVIRONMENT

Earth is an active planet in a constant state of change.
Geological processes continually modify the Earth's surface, destroy old rocks, create new rocks and add to the complexity of ground conditions.

Cycle of geology encompasses all the major processes, which must be cyclic, or they would grind to an inevitable halt.

Land: mainly erosion and rock destruction.
Sea: mainly deposition, forming new sediments.
Underground: new rocks created and deformed.

Earth movements are vital to the cycle; without them the land would be eroded down to just below sea level.
Plate tectonics provide the mechanism for nearly all earth movements (section 09). The hot interior of the Earth is the ultimate energy source which drives all geological processes.

Weathering
Transport
Erosion
Air
Land
Deposition
Sea
Lithification
Earth Movements
Sedimentary Rocks
Igneous Rocks
Metamorphism
Magma
Melting
Metamorphic Rocks

SIGNIFICANCE IN ENGINEERING

Civil engineering works are all carried out on or in the ground. Its properties and processes are therefore significant – both the strengths of rocks and soils, and the erosional and geological processes which subject them to continual change.
Unstable ground does exist. Some ground is not 'terra firma' and may lead to unstable foundations.
Site investigation is where most civil engineers encounter geology. This involves the interpretation of ground conditions (often from minimal evidence), some 3-D thinking, and the recognition of areas of difficult ground or potential geohazards.
Unforeseen ground conditions can still occur, as ground geology can be almost infinitely variable, but they are often unforeseen due to inadequate site investigation.
Civil engineering design can accommodate almost any ground conditions which are correctly assessed and understood.

Geological time is an important concept. Earth is 4000M years old and has evolved continuously towards its present form.
Most rocks encountered by engineers are 10–500M years old. They have been displaced and deformed over time, and some are then exposed at the surface, by erosional removal of rocks that once lay above them.
Underground structures and the ground surface have evolved steadily through geological time.
Most surface landforms visible today have been carved out by erosion within the last few million years, while older landforms have been destroyed.
This time difference is important: the origin of the rocks at the surface may bear no relationship to the present environment. The classic example is Mt Everest, whose summit is limestone, formed in a sea 300M years ago.
Geological time is difficult to comprehend but it must be accepted as the time gaps account for many of the contrasts in ground conditions.

Concepts of scale are important in geology:

Beds of rock extending hundreds of kilometres across country.
Rocks uplifted thousands of metres by earth movements.
Rock structures reaching 1000 m below the ground surface.
Strong limestone crumpled like plasticine by plate tectonics.
Landslides with over 100M tons of falling rock.
Earthquakes a million times more powerful than an atom bomb.
And the millions of years of geological time.

Endless horizontal rocks exposed in Canyonlands, USA

Components of Engineering Geology

The main fields of study:	Sections in this book
Ground materials and structures	02–06
Regional characteristics	09–12
Surface processes and materials	13–18
Ground investigations	07,08,19–23,37
Material properties	24–26,40
Difficult ground conditions	27–36,38,39

Other aspects – fossils and historical geology, mineral deposits and long term processes – are of little direct significance to the engineer and are not specifically covered in this book.

SOME ENGINEERING RESPONSES TO GEOLOGICAL CONDITIONS

Geology	Response
Soft ground and settlement	Foundation design to reduce or redistribute loading
Weak ground and potential failure	Ground improvement or cavity filling; or identify and avoid hazard zone
Unstable slopes and potential sliding	Stabilize or support slopes; or avoid hazard zone
Severe river or coastal erosion	Slow down process with rock or concrete defences (limited scope)
Potential earthquake hazard	Structural design to withstand vibration; avoid unstable ground
Potential volcanic hazard	Delimit and avoid hazard zones; attempt eruption prediction
Rock required as a material	Resource assessment and rock testing

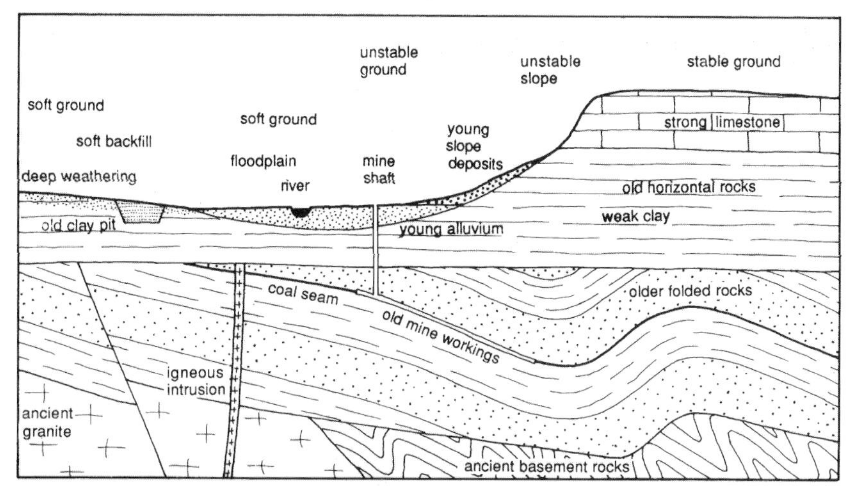

Ground profile through some anonymous region in the English Midlands.

Most rocks were formed 200–300M years ago, when the area was near the equator in a deltaic swamp, disturbed by earth movements then left in a shallow sea.

The ground surface was shaped by erosion in the last million years, when the alluvium and slope deposits partly filled the river-cut valley.

The more difficult ground conditions are provided by the floodplain, soft sediments, deep rockhead, unstable slopes, old mines and the backfilled quarry.

Folded rocks exposed along the South Wales coast

Strong Rocks	Weak Rocks
UCS > 100 MPa	UCS < 10 MPa
Little fracturing	Fractured and bedded
Minimal weathering	Deep weathering
Stable foundations	Settlement problems
Stand in steep faces	Fail on low slopes
Aggregate resource	Require engineering care

STRENGTH OF THE GROUND

Natural ground materials, rocks and soils, cover a great range of strengths: granite is 4000 times stronger than peat soil.

Some variations in rock strength are summarized by contrasting strong and weak rocks in the table.

Assessment of ground conditions must distinguish:

- Intact rock – strength of an unfractured, small block; refer to UCS.
- Rock mass – properties of a large mass of fractured rock in the ground; refer to rock mass classes (section 25).

Note – a strong rock may contain so many fractures in a hillside that the rock mass is weak and unstable.

Ground conditions also vary greatly due to purely local features such as underground cavities, inclined shear surfaces and artificial disturbance.

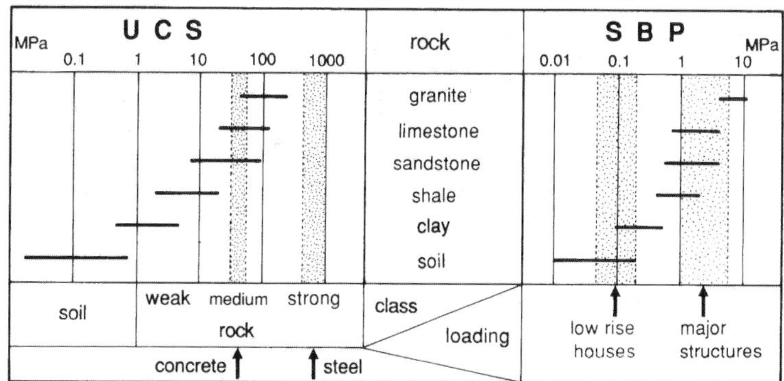

UCS:
Unconfined (or uniaxial) compressive strength: load to cause failure of a cube of the material crushed between two flat plates with no lateral restraint. (Strong and weak limits are simplified; see section 24 for BS criteria.)

SBP:
Safe (or acceptable) bearing pressure: load that may safely be imposed upon rock in the ground: the estimated (or measured) ultimate bearing pressure to fail the rock (allowing for fractures and local zones of weakness) divided by a safety factor between 3 and 5.

ROCKS AND MINERALS

Rocks: mixtures of minerals: variable properties.
Minerals: compounds of elements: fixed properties.

Rock properties broadly depend on:

- strength and stability of constituent minerals;
- interlocking or weaknesses of mineral structure;
- fractures, bedding and larger rock structures.

All rocks fall into one of three families, each with broadly definable origins and properties.

Most rock-forming minerals are silicates – compounds of oxygen, silicon and other elements.

Rock properties can show extreme variations. It is useful to generalize, as in the table below, in order to build an understanding of geology, but it must be accepted that rocks are not engineered materials and their properties do vary from site to site.

For example: most sedimentary rocks are quite weak, and limestone is a sedimentary rock, but some of the limestones are very strong.

Rock family	Igneous	Sedimentary	Metamorphic
Material origin	Crystallized from molten magma	Erosional debris on Earth's surface	Altered by heat and/or pressure
Environment	Underground; and as lava flows	Deposition basins; mainly sea	Mostly deep inside mountain chains
Rock texture	Mosaic of interlocking crystals	Mostly granular and cemented	Mosaic of interlocking crystals
Rock structure	Massive (structureless)	Layered, bedded, bedding planes	Crystal orientation due to pressure
Rock strength	Uniform high strength	Variable low; planar weaknesses	Variable high; planar weaknesses
Major types	Granite, basalt	Sandstone, limestone, clay	Schist, slate

02 Igneous Rocks

Magma is generated by local heating and melting of rocks within the Earth's crust, mostly at depths between 10 and around 100 km. Most compositions of rock melt at temperatures of 800–1200°C. When the magma cools, it solidifies by crystallizing into a mosaic of minerals, to form an igneous rock.

VOLCANIC ERUPTIONS

Eruptions may be violent and explosive if a viscous magma has a high gas pressure, or may be quiet and effusive if the magma is very fluid. There is a continuous range of eruptive styles between the two extremes, and a single volcano may show some variation in the violence of its individual eruptions.

Pyroclastic rocks (meaning fire fragmental) are formed of material, collectively known as tephra, thrown into the air from an explosive volcano. Most tephra is cooled in flight, and lands to form various types of ash, tuff and agglomerate, all with the properties of sedimentary rocks. Some tephra, erupted in turbulent, high-temperature, pyroclastic flows, lands hot and welds into ignimbrite, or welded tuff.

EXTRUSIVE IGNEOUS ROCKS

These form where magma is extruded onto the Earth's surface to create a volcano.
Lava is the name for both molten rock on the surface, and also the solid rock formed when it cools.
Fluid basaltic lavas flow easily to form low-profile shield volcanoes, or near-horizontal sheets of flood basalt.
More viscous lavas, mainly andesitic, build up conical composite, strato-volcanoes, where lava is interbedded with ash and debris, that are thickest close to the vent.

INTRUSIVE IGNEOUS ROCKS

These are formed when magma solidifies below the surface of the Earth. They may later be exposed at the surface when the cover rocks are eroded away.
Batholiths are large blob-shaped intrusions, roughly equidimensional and commonly 5–50 km in diameter. Most are of granite.
Dykes are smaller sheet intrusions formed where magma has flowed into a fissure. Mostly 1–50 m wide; may extend for many kilometres; generally of dolerite. Sills are sheet intrusions parallel to the bedding of the country rocks into which the magma was intruded.

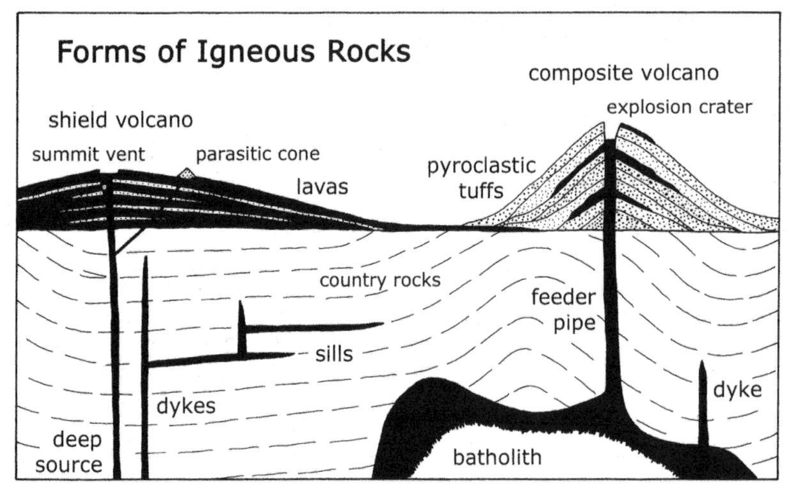

Molten lava ejected from the summit vent of Stromboli volcano.

GRANITE

TYPE	Acid igneous; coarse grained, large scale intrusive (plutonic).
MINERALOGY	Coarse interlocking crystal mosaic with no textural orientation. Quartz 25%, feldspar 50%, micas 15%, mafics 10%.
OCCURRENCE	Large batholiths, exposed at surface by subsequent erosion. Cooled as large bodies 3–15 km beneath surface.
EXAMPLES	Britain: Land's End. USA: Yosemite.
STRUCTURES	Commonly massive and very uniform. Widely spaced sheet jointing, curved due to large exfoliation (caused by cooling and stress relief).
WEATHERING	Slow decay of feldspar to clay, leaving quartz to form sandy soils. Spheroidal weathering leaves rounded corestones in soil matrix.
STRENGTH	High strength with all physical properties good. UCS: 200 MPa. SBP: 10 MPa.
FOUNDATIONS	Very strong rock, except where partially decayed to clay near the surface or along some deep joint zones.
HYDROLOGY	Groundwater only in fractures.
VALUE	Excellent dimension, decorative and armour stone and aggregate.
VARIETIES	Syenite and diorite: have less quartz and are slightly darker. Gabbro: basic, and is much darker. Larvikite: a dark coarse syenite with distinctive internal reflections. Many strong rocks are referred to as granite within the construction trade.

Microscope view, 5 mm across: clear quartz, cloudy feldspar, cleaved mica.

MAIN MINERALS OF IGNEOUS ROCKS

mineral	composition	colour	H	D	common morphology and features
Quartz	SiO_2	clear	7	2·7	mosaic; no cleavage; glassy lustre
Feldspar	$(K,Na,Ca)(Al,Si)_4O_8$	white	6	2·6	mosaic or laths; types – orthoclase and plagioclase
Muscovite	$KAl_2AlSi_3O_{10}(OH)_2$	clear	2½	2·8	splits into thin sheets, due to perfect cleavage } members of the mica group of minerals
Biotite	$K(Mg,Fe)_3AlSi_3O_{10}(OH)_2$	black	2½	2·9	splits into thin sheets, due to perfect cleavage } members of the mica group of minerals
Mafics	Fe–Mg silicates	black	5–6	> 3·0	long/short prisms; hornblende, augite, olivine

Mafic minerals is a convenient term for a group of black silicates whose individual properties are of little significance in the context of most engineering.
Cleavage is the natural splitting of a mineral along parallel planes dictated by weaknesses in the atomic structure.
Mineral strength is a function of hardness and lack of cleavage, along with effects of decay or orientation.

Features are generalized, and exceptions do occur; crystal faces are displayed on museum specimens of most minerals, but are rarely seen in normal rocks.
H = hardness, on a scale of 1–10, from talc the softest mineral of hardness 1, to diamond the hardest of hardness 10. Steel and glass have hardnesses between 6 and 7.
D = density, measured in grams/cm^3 or tonnes/m^3.

CLASSIFICATION OF IGNEOUS ROCKS

This simple classification covers the great majority of igneous rocks. It is based on two parameters which are both significant and recognizable. The main types of igneous rocks can therefore be identified by just colour and grain size.
The form of occurrence determines the structure of the rock in the ground; also, lavas may cool in hours or days while a batholith may take a million years to crystallize, and the cooling rate determines the grain size of the rock.

Chemical composition is determined by what rocks had melted to form the original magma; silica-rich magmas are referred to as acidic (unrelated to pH) and are generally low in iron, so have few black iron minerals, and are therefore lighter in colour than basic rocks.
Porphyritic rocks have scattered larger, older crystals (called phenocrysts) in a finer groundmass.
In fine grained rocks, grains cannot be seen with the naked eye; the limit of 0·1 mm is effectively the same as the limit of 0·06 mm used in soils and sediments.

			occurrence	form	cooling	grain	size
Rhyolite	Andesite	**Basalt**	extrusions	lavas	fast	fine	< 0·1 mm
Porphyry		Dolerite	small intrusions	dykes	medium	medium	0·1–2 mm
Granite	Diorite	Gabbro	large intrusions	batholiths	slow	coarse	< 2 mm

70%		50%	SiO_2 content
acid		basic	classification
viscous		fluid	magma viscosity
explosive		effusive	volcano type
3%		12%	Fe content
10%		50%	mafic minerals
light		dark	colour

Granite and basalt are most abundant because magma viscosity determines the ease of migration. Acid magma is viscous, so most stays in batholiths to form granite, while basic magma is so fluid that most of it escapes to the surface to form basalt lava.

BASALT

TYPE: Basic igneous; fine-grained, extrusive (volcanic).

MINERALOGY: Fine interlocking crystal mosaic with no textural orientation.
May have open vesicles or mineral-filled amygdales (old gas bubbles).
Feldspar 50%, mafics 50%.

OCCURRENCE: Lava flows in bedded sequences. Cooled after flowing from volcano.

EXAMPLES: Britain: Skye and Mull. USA: Columbia Plateau and Hawaii.

STRUCTURES: Sheets or lenses, maybe interbedded with ash or tuff.
Commonly with weathered or vesicular scoria tops on each flow.
Young lavas have smooth pahoehoe or clinkery aa surfaces.
Compact basalt may have columnar jointing (from cooling contraction).

WEATHERING: Rusts and decays to clay soils; maybe spheroidal weathering.

STRENGTH: Compact basalts are very strong.
UCS: 250 MPa. SBP: 10 MPa (less on young lava).

FOUNDATIONS: Variable strength, especially in younger lavas, due to ash beds, scoriaceous or clinkery layers, lava caves and other voids.

HYDROLOGY: Young lavas are generally good aquifers.

VALUE: Good aggregate and valuable roadstone.

VARIETIES: Andesite: intermediate lava, dark or light grey, often weathered red.
Dolerite: medium grained intrusive dyke rock; looks similar to basalt.
Rhyolite: pale grey acid lava, commonly associated with frothy pumice and dense black obsidian glass.

Microscope view, 5 mm across: clear feldspar laths, dark mafics, fine groundmass of same minerals.

03 Surface Processes

Sediment is largely material derived from the weathering of rocks on the Earth's surface (the remainder is mostly organic material).
All rocks weather on exposure to air and water, and slowly break down to form in situ soils.
In most land environments, the soil material is subsequently transported away from its source, and may then be regarded as sediment; this includes the solid debris particles and also material in solution in water.
Natural transport processes are dominated by water, which can sort and selectively deposit its sediment load.
Ultimately all sediment is deposited, mostly in the sea, and mostly as stratified layers or beds of sorted material.
Burial of this loose and unconsolidated sediment, by more layers of material subsequently deposited on top of it, eventually turns it into a sedimentary rock, by the various processes of lithification.

The land is essentially the erosional environment; it is the source of sediment, which forms the temporary soils before being transported away.

The sea is essentially the depositional environment; sediment is buried beneath subsequent layers, and eventually forms most of the sedimentary rocks.

Subsequent earth movements may raise the beds of sedimentary rock above sea level; erosion and removal of the overlying rocks (to form the source material for another generation of sediments and sedimentary rocks) then exposes the old sedimentary rocks in outcrops in a landscape far removed from contemporary seas and in an environment very different from that of the sedimentation.

SEDIMENTARY MATERIALS Most sedimentary rocks are varieties of sandstone, clay or limestone

Mineral grains: mostly quartz, also muscovite (the physically and chemically stable minerals) Rock fragments and volcanic debris (not yet broken down to their constituent minerals)	– SANDSTONES
Breakdown products: clay minerals (formed by reaction of water with feldspar or mafic minerals)	– CLAYS
Organic debris: plant material to form peat and coal (animal soft parts form oil)	– minor rocks
Organic debris: dominated by calcite from marine shell debris Solutes: dominantly calcite precipitated from sea water largely due to biological activity	– LIMESTONES
Solutes: including gypsum and salt, and other less abundant soluble compounds	– minor rocks

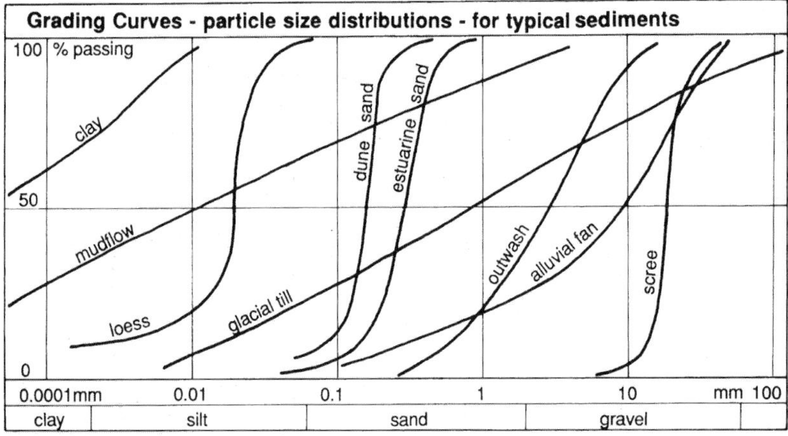

(Mudflow and till are well graded; dune sand and scree are well sorted)

Gully erosion produces new sediment, that builds a steep alluvial fan

SEDIMENT TRANSPORT

The most abundant sediment is clastic or detrital material consisting of particles of clay, sand and rock debris.
Water is far the most important agent of sediment transport. Rivers move the majority of sediment on land. Coarser debris is rolled along the river beds; finer particles are carried in suspension. Water's ability to transport sediment depends on its velocity – larger particles can only be moved by faster flows. Sediment is therefore sorted (to one size) during water transport.
Sediment is also moved in the sea, mainly in coastal waters where wave action reaches the shallow sea bed.
Other transport processes have only limited scope:

- Gravity alone works mainly on the steeper slopes, producing landslides and colluvium.
- Wind moves only fine dry particles.
- Ice transport is powerful, but restricted by climate.
- Volcanoes may blast debris over limited distances.

Some minerals are transported by solution in water.
Organic sediment is rarely carried far from its source.

SEDIMENT DEPOSITION

Water on land Sorted and stratified, mostly sand and clay. Alluvium in river valleys is mostly temporary, later eroded away, except in subsiding deltas. Lake sediment includes salts precipitated due to desert evaporation.

The sea Final destination of most clastic sediment. Sorted and stratified in beds, mostly in shallow shelf seas. Turbidity currents carry sediment into deeper basins. Shell debris in shallow seas, with no land detritus, forms the main limestones.

Slopes Localized poorly sorted scree and slide debris.

Wind Very well sorted sand and silt, mostly in or near dry source areas, so only significant in desert regions.

Ice Unsorted debris dumped in the melt zones of glaciers. Localized today but extensive in past Ice Ages.

Volcanoes Fine, sorted airfall ash, wind-blown over large areas; also coarse unsorted flow and surge deposits, mostly on volcano slopes. Collectively known as pyroclastic sediments (= fire fragmental).

MAIN MINERALS OF SEDIMENTARY ROCKS

Units and terms as for igneous minerals in section 02

mineral	composition	colour	H	D	common morphology and features
Quartz	SiO_2	clear	7	2·7	granular; no cleavage; glassy lustre
Muscovite	$KAl_2AlSi_3O_{10}(OH)_2$	clear	2½	2·8	thin sheets and flakes on perfect cleavage, mica
Kaolinite	$Al_4Si_4O_{10}(OH)_8$	white	the clay minerals		stable type; includes china clay
Illite	$KAl_4AlSi_7O_{20}(OH)_4$	white	maximum crystal size		dominant type, similar to fine muscovite
Smectite	$(Na,Ca)Al_4Si_8O_{20}(OH)_4.nH_2O$	white	only microns across		unstable (variable water); montmorillonite
Calcite	$CaCO_3$	white	3	2·7	mosaic; shell debris; rhombic cleavage on 3 planes
Dolomite	$CaMg(CO_3)_2$	white	3½	2·8	mosaic and rhombs
Gypsum	$CaSO_4.2H_2O$	white	2	2·3	bladed selenite; massive alabaster; fibrous satinspar
Hematite	Fe_2O_3	red	6	5·1	widespread colouring agent
Limonite	FeO.OH	brown	5	3·6	widespread colouring agent; rust, may be yellow
Pyrite	FeS_2	yellow	6	5·0	metallic brassy lustre (fool's gold); common as cubes

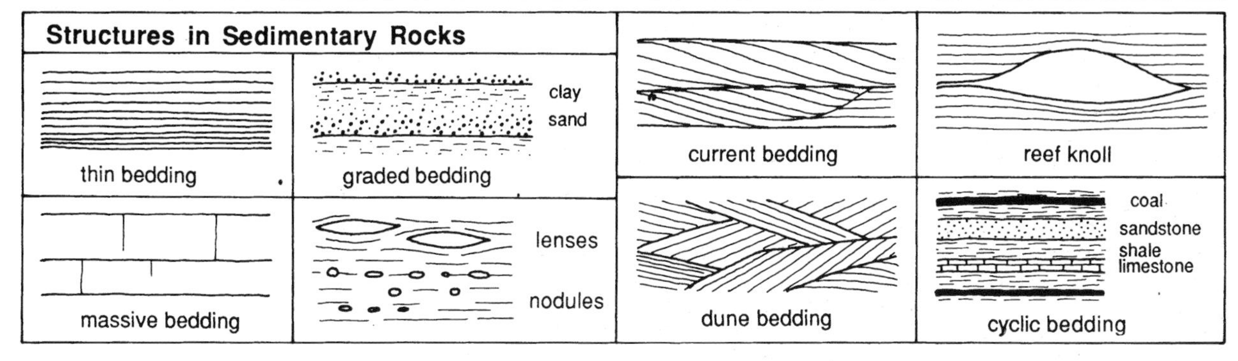

LITHIFICATION

The processes by which a weak loose sediment is turned into a stronger sedimentary rock. Induced by burial pressure and slightly increased temperature beneath a kilometre or more of overlying sediment. The processes of lithification are also known as diagenesis by geologists, referring to the changes which take place after deposition. The results of lithification, notably the increase in strength, are referred to as consolidation by engineers.

Three main processes of lithification:

Cementation The filling of the intergranular pore spaces by deposition of a mineral cement brought in by circulating groundwater. Rock strength is largely dependant on the type of cement, which may be silica (strongest), iron oxides, calcite or clay (weakest). The dominant process in sandstones.

Recrystallization Small scale solution and redeposition of mineral, so that some grains become smaller and some become larger. Result may be similar to cementation, but may produce stronger mosaic texture. Can also include change of state and growth of new more stable minerals. The dominant process in limestones.

Compaction Restructuring and change of grain packing, with decrease in volume, due to burial pressure, with consequent reduction of porosity as water is squeezed out. Increase in strength is due to more grain to grain contact. The dominant process in clays.

CONSOLIDATION generally refers to the increase in strength in clays, due to their restructuring, improved packing, loss of water and reduced porosity caused by compaction under load; it also includes some cementation and new mineral growth.

Normally consolidated clays have never been under a higher load than their existing overburden; these include most clay soils.

Over-consolidated clays have been under a higher load in the past, imposed by cover rocks since removed by erosion; these include nearly all clays within rock sequences. They have lower porosity and higher strength due to their history of burial and exposure.

THE CLAY ROCK CYCLE

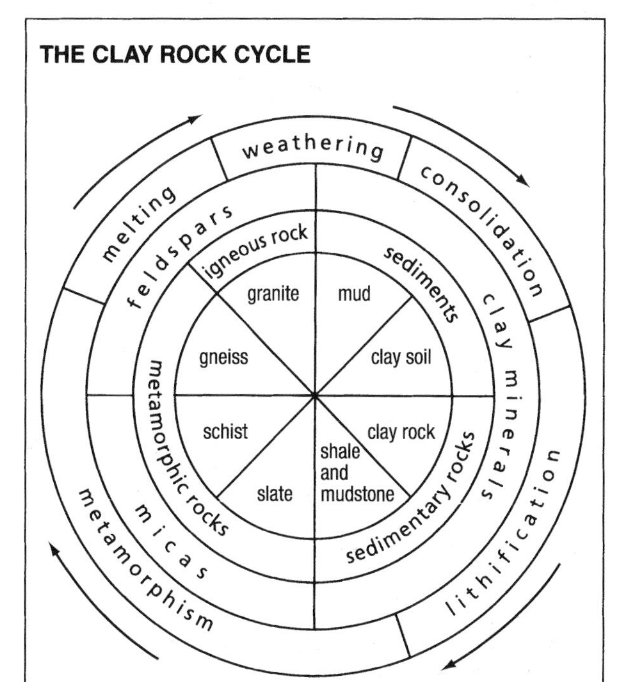

Clay soils and clay rocks related to their metamorphic and igneous derivatives. The eight rocks (and sediment soils) in the core of this cyclic diagram are related by processes (shown in the outer ring) which act in the clockwise direction. Bulk composition is roughly constant, except for the water content which decreases from mud to granite. Only weathering increases the water content, and weathering of any rock may short-circuit the processes by producing mud. Only the main minerals are shown; quartz is present in all the rocks and soils.

(Consolidation also refers to the effect of soil compaction under structural loading, and may be applied specifically to changes taking place when clays are compacted).

04 Sedimentary Rocks

CLASSIFICATION OF SEDIMENTARY ROCKS

A CLASTIC ROCKS		B NON-CLASTIC ROCKS
1. Rudaceous: coarse grained, Conglomerate – rounded fragments Breccia – angular fragments	< 2 mm	1. Carbonates, consisting mainly of calcite Limestone and allied rocks
2. Arenaceous: medium grained, Sandstone and allied rocks	0·06–2 mm	2. Non-carbonates Flint and chert – nodular or banded silica Coal and lignite – lithified peat and plant material Ironstone – any iron-rich sedimentary rock; sand, clay or oolite texture
3. Argillaceous: fine grained, Siltstone – quartz particles Clay and allied rocks	< 0·06 mm	Salt and gypsum – monomineralic rocks deposited by evaporation of water

Electron microscope view of sandstone with quartz grains and weak flaky clay mineral cement. An original partial calcite cement has been removed by weathering.

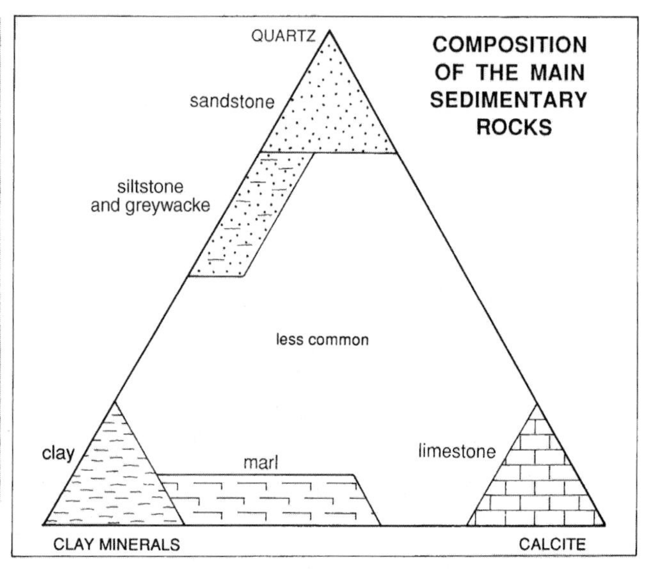

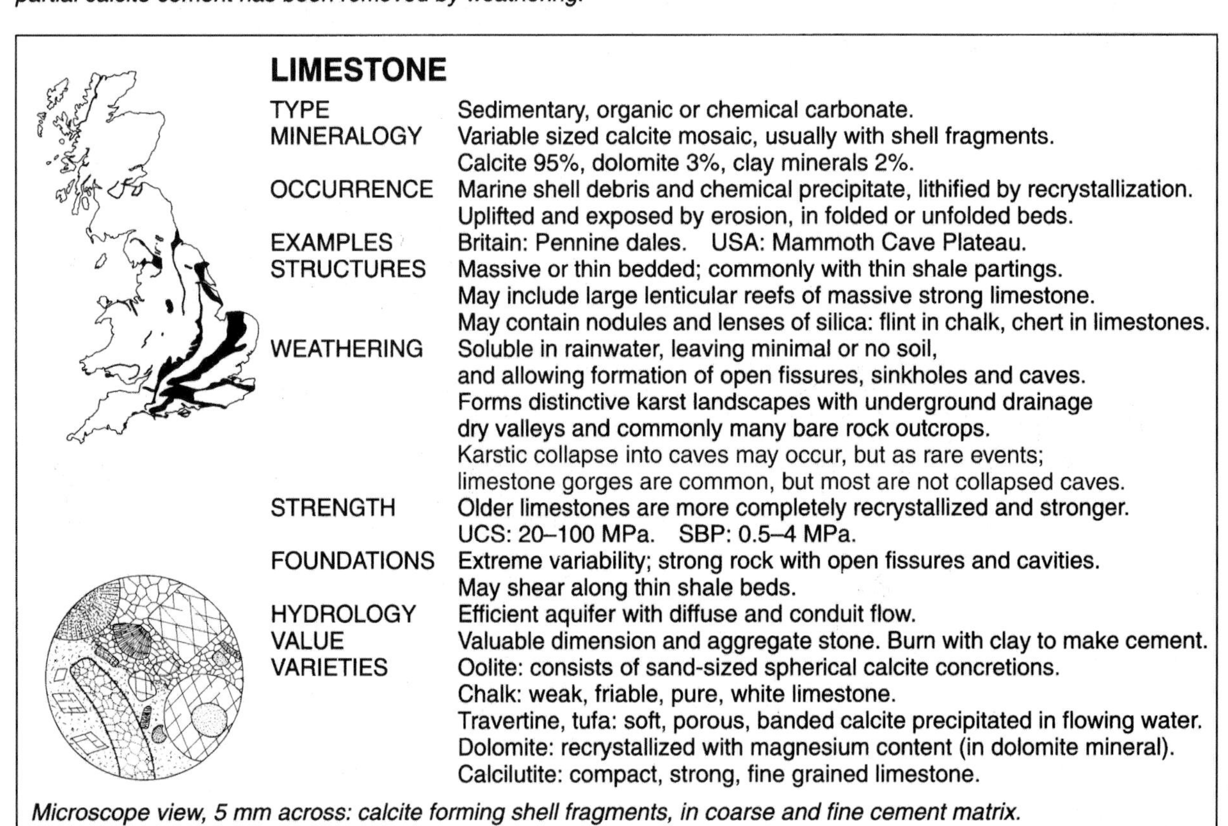

LIMESTONE

TYPE Sedimentary, organic or chemical carbonate.

MINERALOGY Variable sized calcite mosaic, usually with shell fragments.
Calcite 95%, dolomite 3%, clay minerals 2%.

OCCURRENCE Marine shell debris and chemical precipitate, lithified by recrystallization.
Uplifted and exposed by erosion, in folded or unfolded beds.

EXAMPLES Britain: Pennine dales. USA: Mammoth Cave Plateau.

STRUCTURES Massive or thin bedded; commonly with thin shale partings.
May include large lenticular reefs of massive strong limestone.
May contain nodules and lenses of silica: flint in chalk, chert in limestones.

WEATHERING Soluble in rainwater, leaving minimal or no soil,
and allowing formation of open fissures, sinkholes and caves.
Forms distinctive karst landscapes with underground drainage
dry valleys and commonly many bare rock outcrops.
Karstic collapse into caves may occur, but as rare events;
limestone gorges are common, but most are not collapsed caves.

STRENGTH Older limestones are more completely recrystallized and stronger.
UCS: 20–100 MPa. SBP: 0.5–4 MPa.

FOUNDATIONS Extreme variability; strong rock with open fissures and cavities.
May shear along thin shale beds.

HYDROLOGY Efficient aquifer with diffuse and conduit flow.

VALUE Valuable dimension and aggregate stone. Burn with clay to make cement.

VARIETIES Oolite: consists of sand-sized spherical calcite concretions.
Chalk: weak, friable, pure, white limestone.
Travertine, tufa: soft, porous, banded calcite precipitated in flowing water.
Dolomite: recrystallized with magnesium content (in dolomite mineral).
Calcilutite: compact, strong, fine grained limestone.

Microscope view, 5 mm across: calcite forming shell fragments, in coarse and fine cement matrix.

SANDSTONE

TYPE	Sedimentary, clastic, arenaceous
MINERALOGY	Medium grained, with sand grains mostly of quartz, set in cement of quartz, calcite, clay or other mineral. Quartz 80%, clay minerals 10%, others 10%.
OCCURRENCE	Sand of marine, river or desert origin, lithified by cementation. Uplifted and exposed by erosion, in folded or unfolded beds.
EXAMPLES	Britain: Pennine moors and edges. USA: Canyonlands.
STRUCTURES	Massive or thin bedded; commonly interbedded with shale. May have cross bedding inherited from deltaic or dune origin.
WEATHERING	Crumbles to sand, forming sandy well-drained soils.
STRENGTH	Older sandstones tend to be better cemented and stronger. Clay cements are notably weak; quartz cements are generally strong. UCS: 10–90 MPa SBP: 1–4 MPa.
FOUNDATIONS	Generally strong material, unless poorly cemented or with weak cement.
HYDROLOGY	Productive aquifer with diffuse flow.
VALUE	Most sandstones abrade too easily for use as aggregate; some may yield good dimension stone.
VARIETIES	Flagstone: thinly bedded due to partings rich in mica flakes. Grit: imprecise colloquial term for strong sandstone. Greywacke: old, partly metamorphosed, strong; interbedded with slate. Flysch: young and weak; interbedded with shale or clay. Tuff: volcanic ash of sand grain size; lithified or unlithified.

Microscope view, 5 mm across: mostly quartz grains, two cement types.

Eroded remnants of once continuous beds of sandstone in Monument Valley, northern Arizona, USA. A massively bedded sandstone forms the vertical sided buttes, and overlies a thinly bedded sandstone with many shale layers.

CLAY

TYPE	Sedimentary, clastic, argillaceous.
MINERALOGY	Fine grained structureless mass of clay minerals, commonly with a proportion of small silt grains of quartz. Illite 60%, kaolinite 20%, smectite 10%, others 10%.
OCCURRENCE	Mud, mainly of marine origin, lithified by compaction and water expulsion. Uplifted and exposed by erosion, in folded or unfolded beds.
EXAMPLES	Britain: London Clay. USA: Dakota Badlands.
STRUCTURES	Commonly featureless and unbedded, but may be bedded with variable silt and organic content. May have nodules (hard rounded lumps) with stronger mineral cement.
WEATHERING	Reverts to mud, forming heavy clay soils.
STRENGTH	Older, more lithified and unweathered clays have higher strength. Younger clays have properties transitional to those of low strength soils. UCS: 1–20 MPa. SBP: 0.1–1 MPa.
FOUNDATIONS	Weak material with low, variable strength related largely to water content; prone to slow creep and plastic deformation; high potential compaction may cause high and differential settlement under structural load.
HYDROLOGY	Aquiclude
VALUE	Watertight fill, bricks, cement.
VARIETIES	Mudstone: more lithified, massive and stronger. Shale: more lithified, laminated and fissile. Marl: clay or mudstone with significant calcite content. Siltstone: mainly quartz grains, essentially a fine grained sandstone.

Microscope view, 5 mm across: clay groundmass, silty layers.

05 Metamorphic Rocks

Metamorphic rocks are created by changes induced at high temperature (up to about 600°C) and/or high pressures (around 500 MPa at 20 km depth). These changes (metamorphism) take place in the solid state. The type of metamorphic rock produced depends on the original rock material that was metamorphosed and the temperature and pressure conditions which were imposed.

METAMORPHIC CHANGES IN ROCK

Recrystallization forms a strong mineral mosaic, notably in marble.

New minerals grow at the expense of less stable minerals in the new conditions of high temperature and pressure. Most important changes are clay minerals → micas → feldspars and mafics. Micas are the most significant minerals in metamorphic rocks and only change to feldspars at the highest grade of metamorphism.
A green colour is typical of low grade metamorphic rocks that contain significant chlorite and epidote.

Directional pressure within the solid state creates mineral orientation within the regionally metamorphosed rocks. New minerals grow in the line of least resistance – perpendicular to the maximum pressure – to cause foliation, or banding, within these rocks.

Planar weaknesses in the foliated metamorphic rocks are created by the parallel micas splitting along their mineral cleavage – causing rock cleavage (also known as slaty cleavage) and schistosity – both of which are independent of any original bedding.

Non-foliated metamorphic rocks have stronger isotropic structure. These include hornfels, formed by thermal metamorphism of clay without high pressure; also marble, and gneiss with little or no mica.

TYPES OF METAMORPHISM

Regional metamorphism involves high temperature and pressure. Occurs in mountain chains due to continental collision on plate boundaries. Extends over large areas.
Thermal or **contact metamorphism** involves high temperature only. Occurs in metamorphic aureoles, each 0·001–2 km wide, around igneous intrusions where rock has been baked.
Dynamic metamorphism at high pressure only is rare.

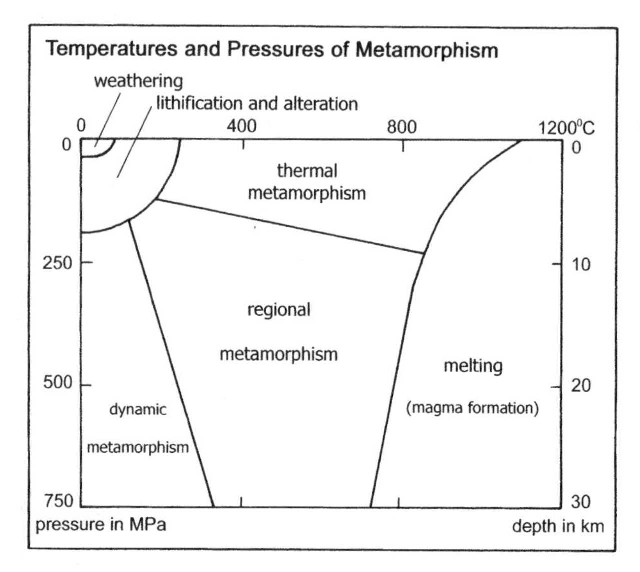

Grade of metamorphism is the overal extent of change, notably in the sequence (within regional metamorphism) from slate to schist to gneiss. Sequence of changes can be seen in the rock cycle diagram in section 03.

METAMORPHISM OF DIFFERENT ROCKS

Limestone → marble: by recrystallization of calcite, forming strong mosaics.
Sandstone → quartzite: by recrystallization of quartz, forming very strong mosaics.
Basalt → greenstone: by limited growth of new green minerals.
Granite shows little change: stable in metamorphic conditions.
Clay (and rock mixtures) → hornfels, slate, schist or gneiss: depending on type and grade of metamorphism.

MAIN METAMORPHIC ROCKS Derived from clay or mixtures of rocks

name	grain size	main minerals	structure	strength	UCS (MPa)
Hornfels	fine	mica, quartz, clay minerals	uniform	very strong	200
Slate	fine	mica, quartz, clay minerals	cleavage	low shear, high flexural	20–120
Schist	coarse	mica, quartz	schistosity	very low shear	20–70
Gneiss	coarse	quartz, feldspar, mafics, mica	foliation	strong	100

ROCK ALTERATION

Alteration includes various processes which affect rocks, usually involving water at lower temperatures and pressures than metamorphism.
Weathering involves rainfall water coming from above; a near-surface feature (see section 13).
Hydrothermal alteration involves hot water rising from below, commonly from volcanic source.
Metasomatism involves chemical replacement by elements carried in solution.

Alteration is commonly localized within a few metres of major faults or fractures; it may occur throughout zones a kilometre or more across.

New hydrated, weak minerals are the main product of alteration processes; normally the altered rock is therefore significantly weaker than the original.
Alteration may be indicated by local colour changes, notably green or yellow.

Chloritization: very low grade metamorphic growth of weak, green chlorite.
Kaolinization: alteration of feldspars to kaolinite (clay mineral).
Sericitization: alteration of feldspars to sericite flakes (similar to fine grained muscovite).
Iron alteration: rusting and decay of iron minerals to yellow or brown limonite.

MAIN MINERALS OF METAMORPHIC ROCKS

mineral	composition	colour	H	D	common morphology and features
Quartz	SiO_2	clear/white	7	2·7	mosaic; no cleavage; glassy lustre
Feldspar	$(K,Na,Ca)(Al,Si)_4O_8$	white	6	2·6	mosaic or short prisms
Muscovite	$KAl_2AlSi_3O_{10}(OH)_2$	clear	2½	2·8	thin sheets and flakes; perfect cleavage }
Biotite	$K(Mg,Fe)_3AlSi_3O_{10}(OH)_2$	black	2½	2·9	members of the mica group of minerals }
Chlorite	$Mg_5Al_2Si_3O_{10}(OH)_8$	blue-green	2	2·7	small flakes; perfect cleavage
Epidote	$Ca_2(Al,Fe)_3Si_3O_{12}.OH$	green	6	3·3	small laths
Calcite	$CaCO_3$	white	3	2·7	mosaic; rhombohedral cleavage on 3 planes
Kaolinite	$Al_4Si_4O_{10}(OH)_8$	white	2	2·6	fine powdery clay mineral
Limonite	$FeO.OH$	brown	5	3·6	rusty staining

Other metamorphic minerals, such as hornblende, garnet and andalusite, may be present, but have little influence on rock properties. Units and terms as explained for igneous minerals in section 02.

SCHIST

TYPE Regional metamorphic, medium grade, foliated.
MINERALOGY Coarse grained mosaic with banding and conspicuous parallel orientation. Micas 35%, chlorite 20%, quartz 25%, others 20%.
OCCURRENCE Regional metamorphism of clays and mixed rocks at high temperature and pressure, in structurally complex cores of mountain belts on convergent plate boundaries.
EXAMPLES Britain: Scottish Highlands. USA: Inner gorge of Grand Canyon.
STRUCTURES Prominent schistosity due to parallelism of abundant mica, commonly with foliation banding and complex folding and crumpling.
WEATHERING Slow alteration to clays.
STRENGTH Anisotropic: compressive strength varies by factor of 5 across or oblique to schistosity. Very low shear strength; weakest with higher chlorite or mica content. UCS: 20–70 MPa. SBP: 1–3 MPa.
FOUNDATIONS Commonly weak, easily sheared.
HYDROLOGY Aquiclude.
VALUE Minimal.
VARIETIES Slate: finer grained, with excellent rock cleavage. Phyllite: intermediate between slate and schist. Gneiss: less mica, more quartz, and higher strength.

Microscope view, 5 mm across: subparallel mica flakes, patches of quartz mosaic.

HORNFELS

TYPE Thermal metamorphic, non-foliated.
MINERALOGY Fine grained mosaic, with no mineral orientation or foliation. Micas 30%, quartz 30%, others 40%.
OCCURRENCE Thermal (contact) metamorphism of clay at high temperature, in metamorphic aureoles up to 1 km wide around major igneous intrusions.
EXAMPLES Britain: Dartmoor margins. USA: Sierra Nevada margins.
STRUCTURES Commonly closely jointed with sharp fractures and local irregularities. May have inherited structures from original rock.
WEATHERING Very slow alteration to clays.
STRENGTH Fine grained materials are generally very strong. UCS: 250 MPa. SBP: 4 MPa.
FOUNDATIONS Strong rock.
HYDROLOGY Aquiclude.
VALUE Good aggregate stone of high strength and low abrasion.
VARIETIES Marble: metamorphosed limestone, consists of recrystallized calcite. Quartzite: metamorphosed sandstone, consists of recrystallized quartz. Greenstone: metamorphosed basalt, with some new green minerals.

Microscope view, 5 mm across: fine groundmass of quartz and mica; large andalusite and mica flakes.

06 Geological Structures

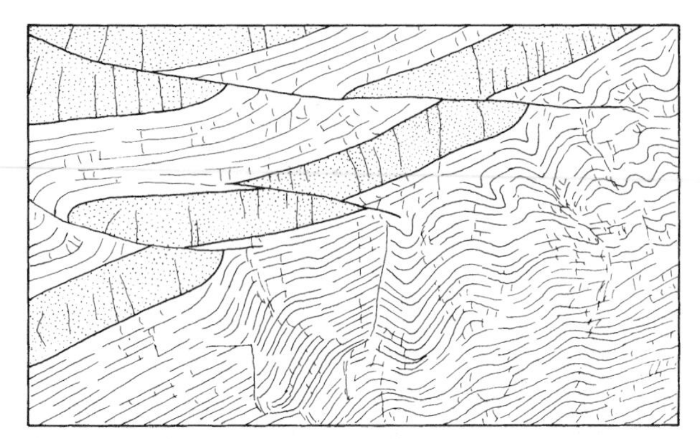

Strongly folded and faulted rock in Greek road cut.

Earth movements involve plastic folding and brittle fracture of rocks, as well as uplift and subsidence. These are tectonic features, caused by large scale movements of crustal plates (section 09). Under the high confining pressures at kilometres of depth, and over the long time scales of tectonic processes, most rocks may show plastic deformation, and fractures occur when and where the plastic limits are exceeded.

Outcrop is an exposure of rock at the surface (or the area of a rock lying directly beneath a soil cover).

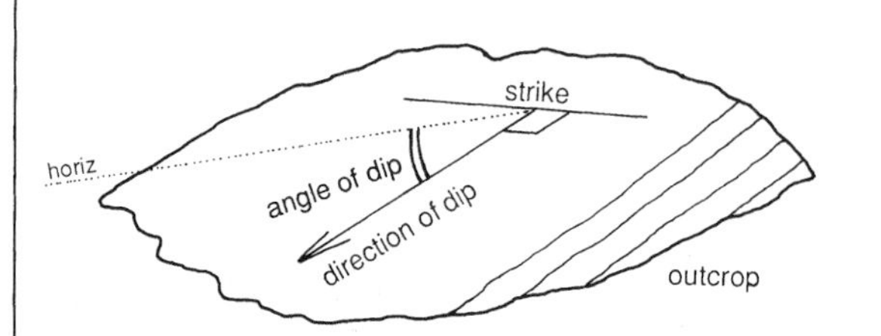

Dip is the angle in degrees below the horizontal.
Direction of dip is down the dip.
Strike is direction of horizontal line on a dipping surface.
These refer to bedding or any geological structures.
Rock dip is used to avoid confusion with ground slope.

FRACTURES

Faults are fractures which have had displacement of the rocks along them.
Throw is the vertical component of fault displacements.
Faults are described by reference to their downthrow side; this is relative movement and may be due to the other side having moved up.

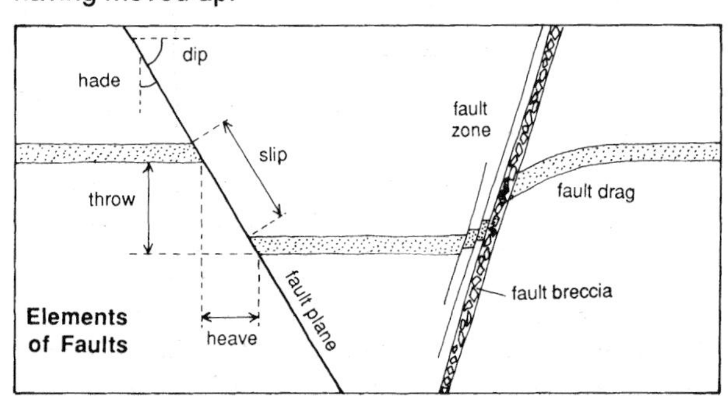

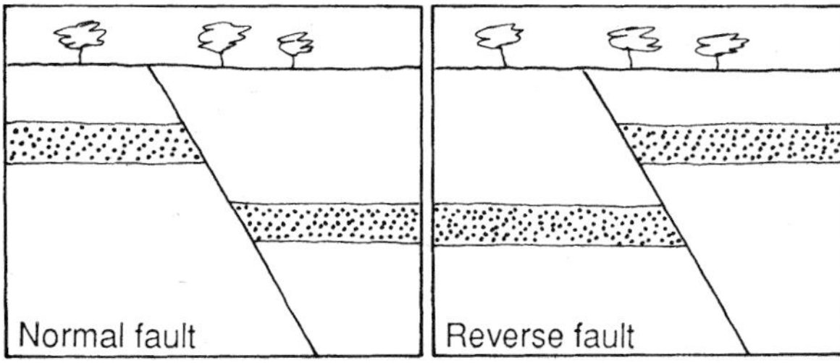

Joints are rock fractures with no movement along them. They are formed by tectonic stressing and are developed in nearly all rocks.
Joint densities and individual joint lengths are infinitely variable.
Groups of sub parallel joints form joint systems.
The dominant fractures within sedimentary rocks are usually the bedding planes. Many bedding planes are very thin bands or partings of shale or clay between units of stronger rocks. Others are clean breaks, or joints, developed tectonicly along the slightest of contrasts within the deposition sequence.
Slaty cleavage and schistosity are also effectively types of joints.
All joints are structural weaknesses, whose density, extent and orientation are major influences on rock mass strength (section 25).
Massive rocks have less fractures, joints or structural weaknesses.

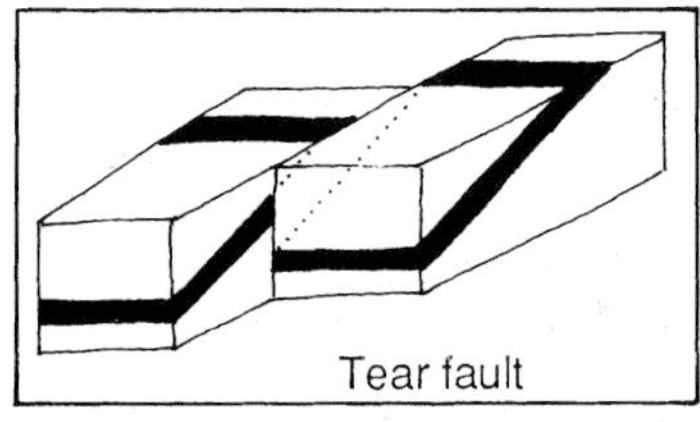

Fault types are recognized by relationship of downthrow to dip of the fault plane.

- Normal faults form under tension; downthrow is on downdip side.
- Reverse faults form under compression; downthrow side is opposite to dip.
- Vertical faults are not easily distinguished as normal or reverse.
- Thrust faults or thrusts, are reverse faults with low angles of dip.
- Tear faults have horizontal displacement (with apparent throw in dipping rocks).
- Grabens are downfaulted blocks between two normal faults.

FEATURES OF FAULTS

Faults commonly create zones of broken ground – weaker and less stable than the adjacent rock – with implications for foundation bearing capacity, slope stability and tunnel roof integrity.
Sudden movements along faults (when tectonic stresses accumulate to overcome frictional resistance) cause earthquakes – vibrations transmitted through the surrounding ground (section 10).
Old faults (including all those in Britain) cannot displace ground surface which has evolved subsequent to any fault movement. Fault line scarps and valleys may appear in a landscape due to differential erosion across the fault zone and adjacent contrasting rocks.

Fault breccia: coarse angular broken rock debris in zone (0·1 –100 m wide) along fault; commonly a zone of groundwater flow.
Fault gouge: finely ground rock paste in thin zone along fault plane.
Fault drag: disturbance and folding of rock near fault.
Slickensides: scratches and polishing on fault planes, and on bedding plane faults within tight folds.
Veins: sheets of mineral infill deposited by hydrothermal water in fractures or fissures in rock. They occur in joints or faults. Most veins are of quartz or calcite – white streaks in rock faces. Larger veins (most on faults) can contain valuable minerals – may have been mined out.

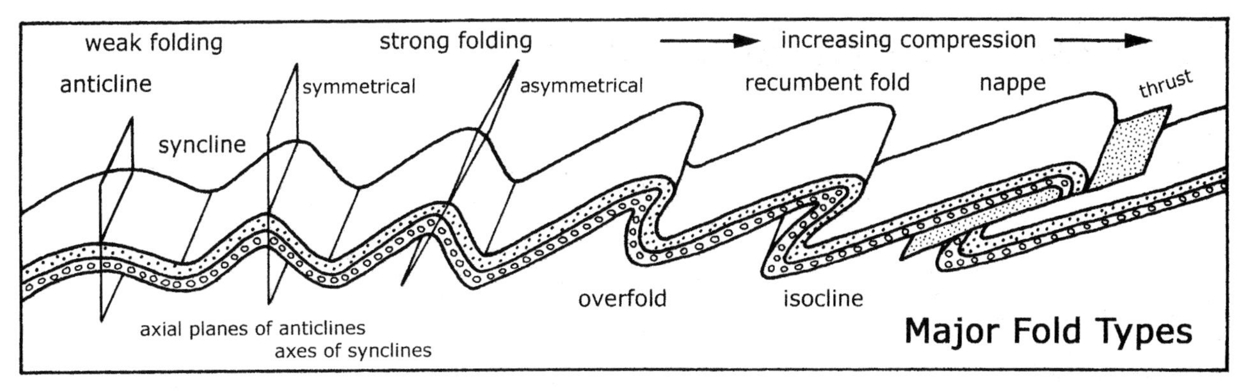

FOLDS

Folds are upward anticlines or downward synclines.

They may be gentle, moderate, or strong.

Folds may be rounded or angular.

Overfolds and recumbent folds have dips past vertical. Isoclines have parallel dips on both sides.
Nappes are recumbent folds sheared along the central line with the development of a thrust fault, usually with large displacement.

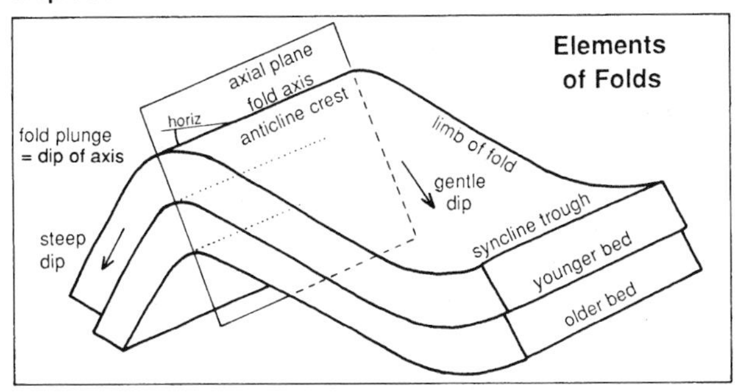

Escarpments, or cuestas, are asymmetrical hills of dipping beds of strong rock, exposed by differential erosion of weaker rocks above and below.

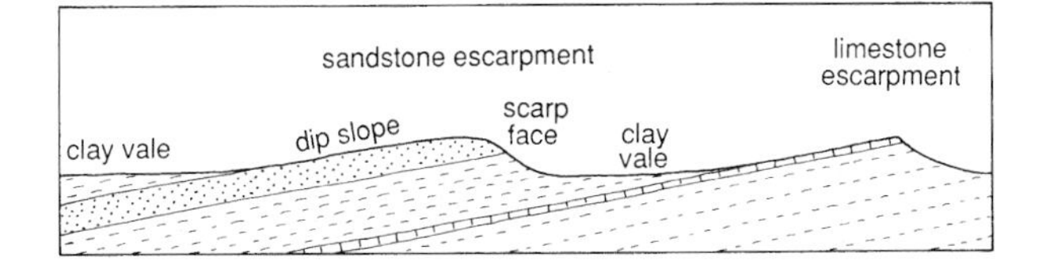

Succession of rocks Older rocks generally lie below younger rocks, and are only exposed by erosion. Reference to old and young rocks avoids confusion with high and low outcrops referring to topographical position.

Inlier is an outcrop of old rocks surrounded by the outcrops of younger rocks; its presence on a map indicates either an eroded anticline or a valley.

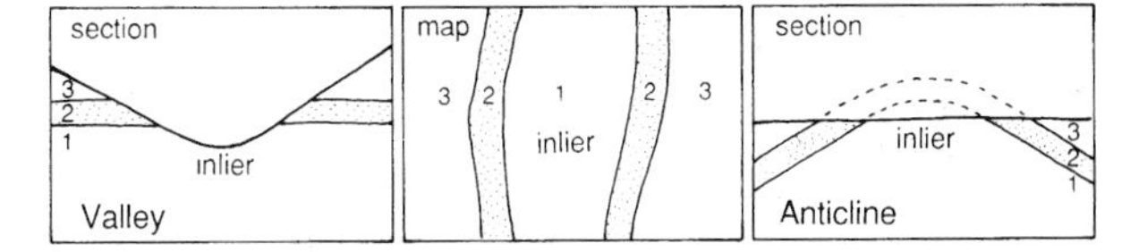

Outlier is an outcrop of young rocks surrounded by old, due to either an eroded syncline or a hill.

Unconformity is the plane or break between two sequences of rocks with different dips. It indicates a period of earth movements and tectonic deformation between the times of sediment deposition. It forms a major structural break – the older rocks must be more lithified and folded, and perhaps more metamorphosed, than the younger rocks above the unconformity.

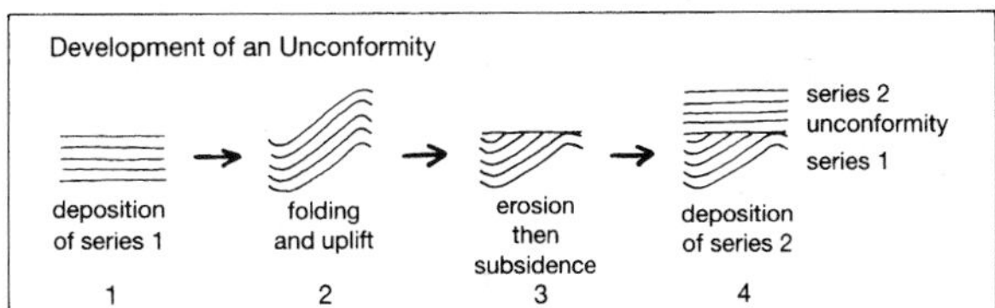

NON-TECTONIC STRUCTURES

Localized structures formed in shallow rocks and soils, by erosion processes and shallow ground deformation, unrelated to regional tectonic structures.
Camber folds develop in level or low-dip rocks where a clay (or soft shale) underlies a strong sandstone or limestone. The clay is plasticly squeezed out from beneath the hill due to the differential loads upon it. Valley bulge is the floor lift (eroded away) and the structural disturbance left beneath it. Most clay is squeezed out from close to the valley side (or scarp edge), so that overlying stronger rocks sag and camber towards the valley.
Gulls are open or soil-filled fissures in the strong rocks of cambered valley sides, opened by camber rotation and perhaps also by sliding.

Post-glacial cambered ground, or foundered strata, is common in the sedimentary rocks of England; it causes fissured rock masses and potential landslides along many valley sides and scarp faces.

Glacial drag: shallow local disturbance, with folds, overfolds and faults in soils and weak rocks overridden by Pleistocene glaciers.
Unloading joints: stress-relief fractures close to and parallel to ground surface due to erosional removal of overburden cover rocks.
Landslip fissures: open fissure and normal faults developed in head zones of slopes prior to failure.
Contraction joints: cooling joints in igneous rocks, including columnar basalt.

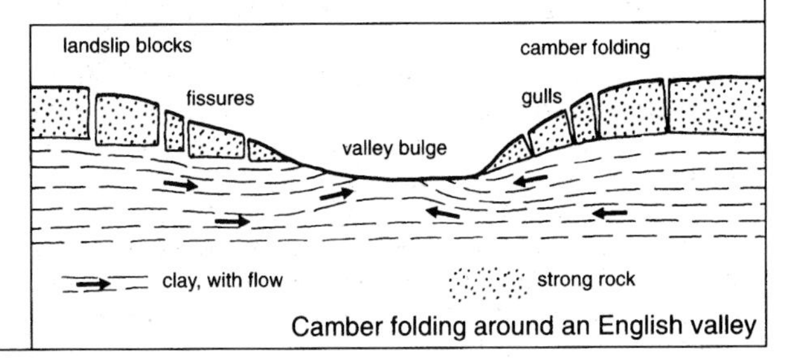

Camber folding around an English valley

07 Geological Maps and Sections

Geological maps show outcrops (where the rocks meet the surface). Shapes of outcrops depend on the shape at the surface and the shape of the rock structure. Surface shape is known (from topographic contours): therefore rock structure can be interpreted.
An important rule: where more than one interpretation is possible, the simplest is usually correct.
Map interpretation is therefore logical and straightforward if approached systematically. Maps remain the best way of depicting 3-D rock structure on a piece of paper.

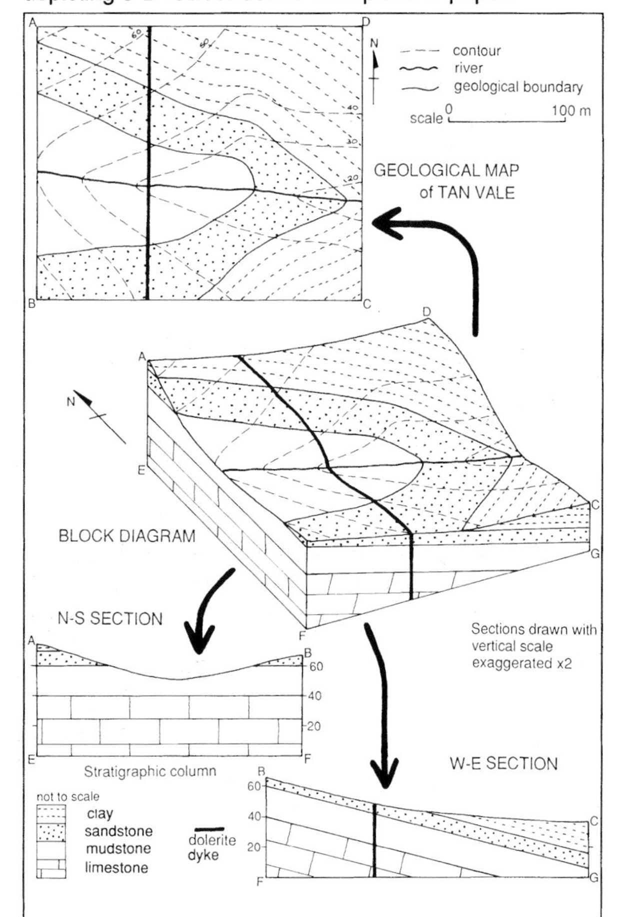

Relationships between a geological map, geological cross-sections and the three-dimensional structure. The north–south section is drawn along the strike, and therefore does not demonstrate the dipping geological structure.

STAGES OF MAP INTERPRETATION

1. Identify faults and unconformities (structural breaks).
2. Identify dips by V in Valley Rule.
3. Determine succession (unless already given).
4. Identify fold axes (from dips and outcrop bends).
5. Draw stratum contours (if detail is needed).
6. Draw cross-section to show sub-surface structure.

GEOLOGICAL MAP SYMBOLS

- →4 dip, direction, and amount in degrees
- horizontal beds; vertical beds
- fault, tick on downthrow side
- anticline; syncline
- sandstone; shale or clay
- limestone; igneous rock
- alluvium; till

OUTCROP PATTERNS

Six basic concepts cover all outcrop patterns, and enable most geological maps to be interpreted successfully.
Horizontal beds have outcrops which follow the contours because they are at constant altitude (limestone on the Scar Hill map).
Vertical beds have straight outcrops which ignore the contours (the dyke on the Tan Vale map).
Dipping beds have curved outcrops which cut across and respond to the contours because outcrops shift downdip as erosion lowers the surface (sandstone on both maps).
Dip direction is recognized by the ***V in Valley Rule***: an outcrop of a dipping rock bends round a V shape where it crosses a valley, and the V of the outcrop points (like an arrowhead) in the direction of dip, regardless of the direction of valley slope and drainage.
This works because the outcrop is shifted furthest downdip at its lowest point where it crosses the valley floor (see the Tan Vale map and diagram).
(The rule does not apply in areas of low dip, where outcrops nearly follow contours, so point upstream.)
On level ground, dipping beds have straight outcrops along the direction of strike.
Succession is recognized by younger rocks coming to outcrop in direction of dip. Conversely, if succession is known, the dip is in the direction of younger outcrops – the easiest way to recognize dip on most maps.

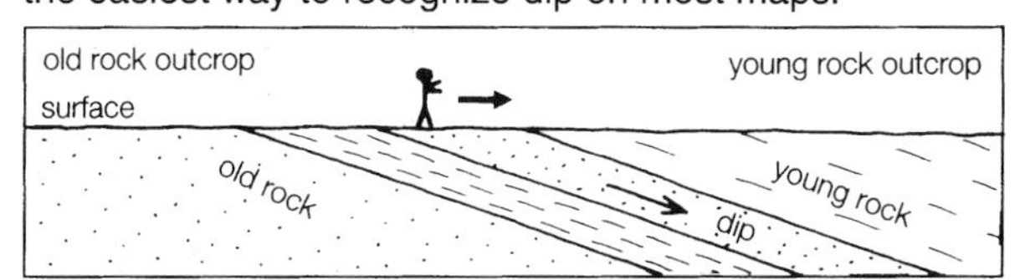

Width of outcrop is greater on thicker beds and at lower dips.

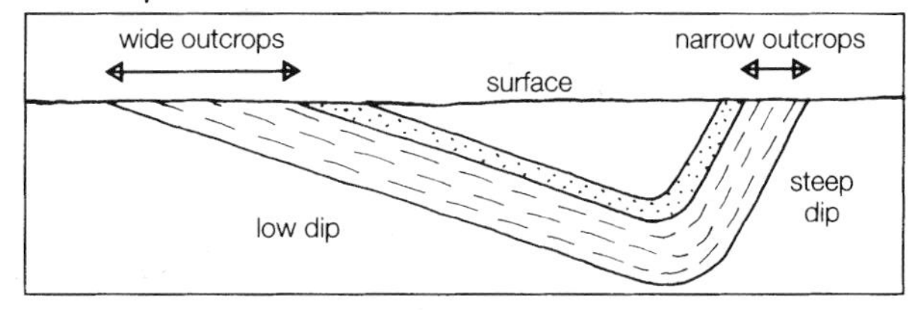

RECOGNITION OF STRUCTURES

Unconformity is recognized where one outcrop (of a younger bed) cuts across the ends of outcrops of older beds, as does the limestone on the Scar Hill map.
Faults are usually marked and keyed on maps. They may cut out, offset or repeat outcrops of beds. Fault dip is recognized by V in Valley Rule. Downthrow side of a fault is the side with younger outcrop because the older rocks have been downthrown to beneath surface level.

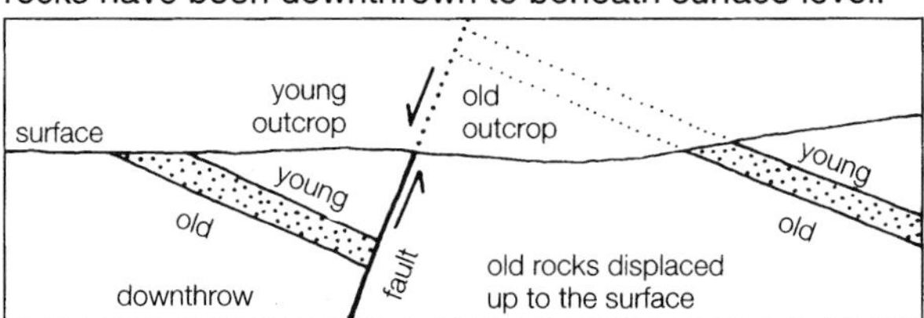

Folds are recognized by changes in dip direction, and also by outliers and inliers not due to topography. Most important, folds are recognized by bends in outcrop: any outcrop bend must be due to either a fold or a topographic ridge or valley. Each outcrop bend should be interpreted, as on the Scar Hill map.

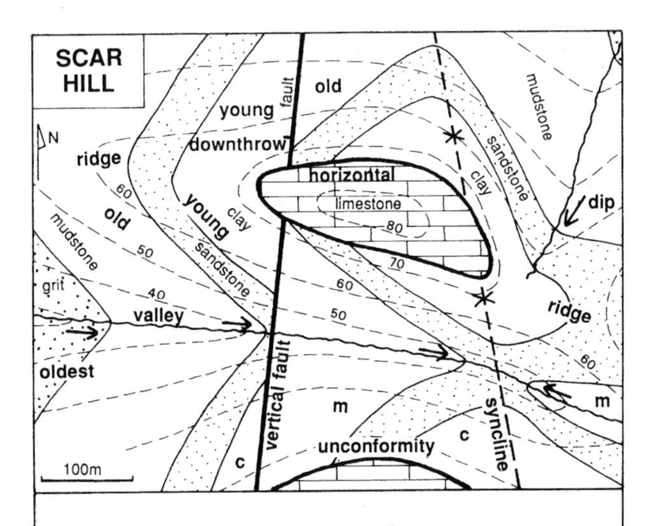

Interpretation of the Scar Hill map, using bends in outcrop to identify either topographical features, direction of dip where crossing a valley, or fold axes if not explained by any topographical feature.

STRATUM CONTOURS

These are lines drawn on a map joining points of equal height on a bed (or stratum). They are like topographic contours, except that they show the shape of buried geological structures. Each contour is labelled with its altitude and the bed boundary to which it refers.

They are drawn by joining points of known equal height on one geological boundary – where its outcrop crosses a topographic contour. The surface information of the map is therefore used to construct the stratum contours, which provide data on the underground geology.

With uniform dip, stratum contours are straight, parallel and equidistant.

Stratum contours have been drawn on part of the Tan Vale map, lower down this column:

- they extend right across the map;
- some apply to two boundaries and are double labelled;
- every boundary/contour intersection has a stratum contour drawn through it;
- labels refer to the base of a bed.

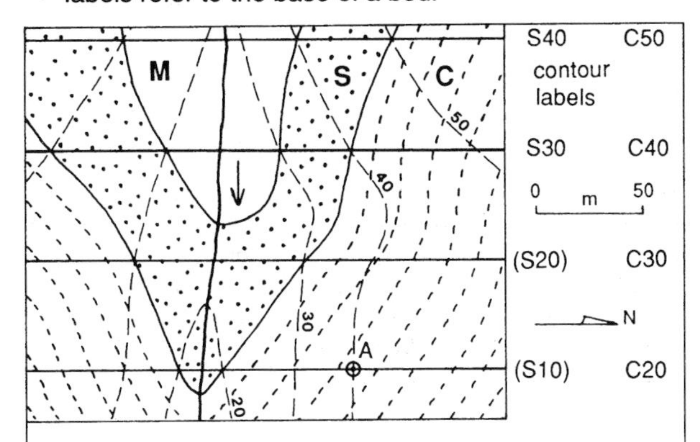

Stratum contours on part of the Tan Vale map

Information can be read from the stratum contours:

- Dip direction is east, 090 (90° from contours);
- Dip amount is 1 in 5 (10 m contours are 50 m apart);
- Sandstone has vertical thickness of 10 m (sandstone base 30 is same contour as clay base 40);
- True thickness = vertical thickness x cosine dip;
- Depths to any rock can be read off at any point.

The stratum contours indicate that a borehole at point A would pass through 20 m of clay, then reach the sandstone which would continue to a depth of 30 m, below which lies the mudstone.

DRAWING A CROSS-SECTION

A cross-section is drawn by projecting the data from a single line on the map onto a profile of the same scale (or with vertical exaggeration if required).

The topography and each geological boundary are constructed individually from the relevant contours, whose intersections on the section line are projected to their correct height on the profile.

Three stages in drawing a profile across the Scar Hill map are shown below. The projection lines and ringed points are only included to demonstrate stages 1 and 2.

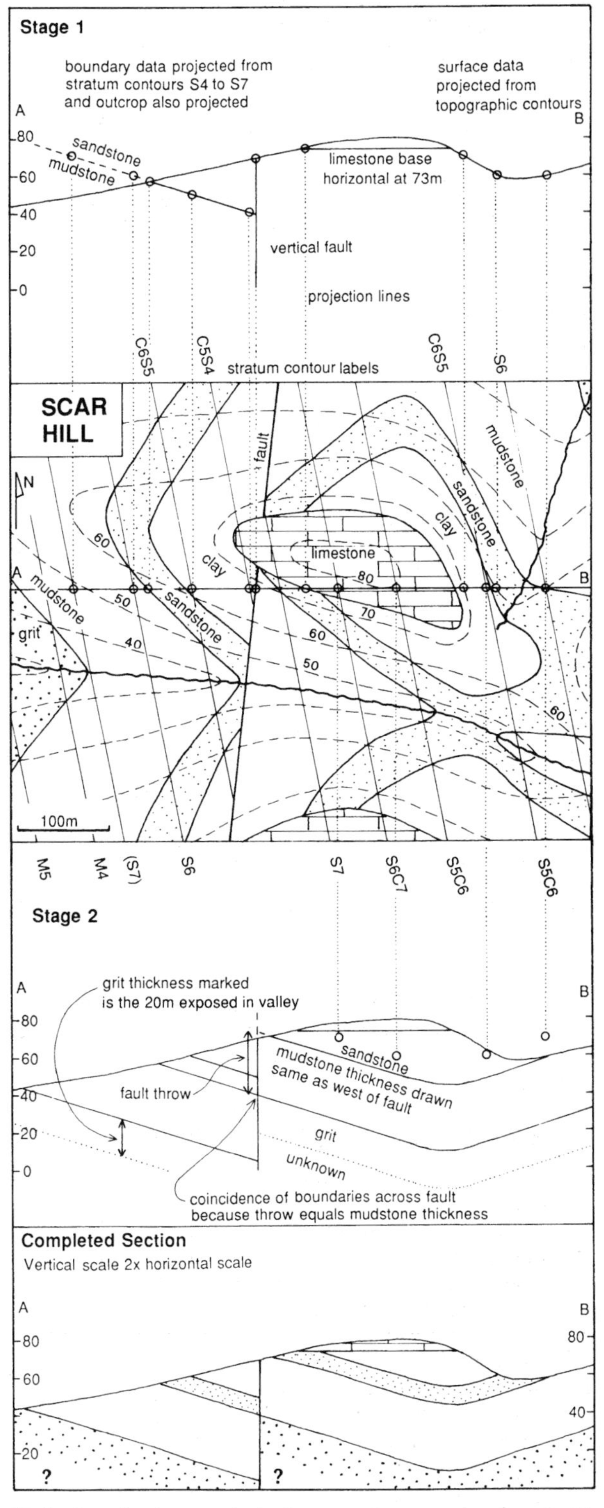

Note two features of stratum contours only shown on the Scar Hill map: they do not cross the fault; they do not relate to the limestone above the unconformity.

08 Geological Map Interpretation

Most published maps have scales between 1:10 000 and 1:100 000.
Low relief cannot be shown accurately at these scales, and therefore stratum contours cannot be drawn to show the geology.
Principles of outcrop shape, bed relationships and structure recognition (from section 07) still apply, but interpretation and section drawing cannot rely on stratum contours.

SOME BASIC CONCEPTS

Most outcrops are laterally uniform sedimentary rocks.
These are in parallel beds of roughly constant thickness.
They are folded and crumpled into parallel curves.
Bed thinning and splitting is rarely seen in small areas.
Beds do not form patternless wedges and blocks.
Heavily faulted areas can provide local complexity.
Intrusions and volcanic rocks have more varied shapes.
Drift sediments form thin, but variable, surface layers.

Geological Map of Oakunder

This is a typical example of a geological map: it shows all rock outcrops, and has some dip arrows; stratigraphic column shows succession and bed thicknesses; topography is only shown by river valleys.

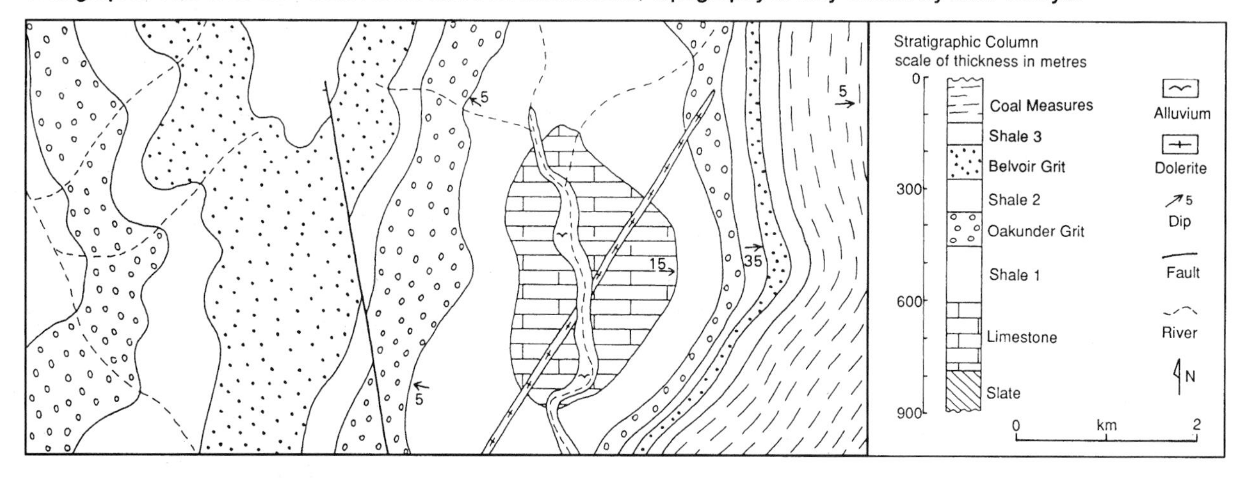

INITIAL INTERPRETATION

Follow stages 1–4 (in section 07).
Add interpretation data to map.
Dips from V in Valley rule.
Dips from succession rule.
Outcrop widths indicate dip.
Width/thickness = dip gradient.
Three shale beds distinguished: numbered in stratigraphic column, and labelled on map.
Fault and dyke straight, so vertical.
Fault is minor – dies out to north, has small outcrop displacements
Folds mirror outcrop sequences.

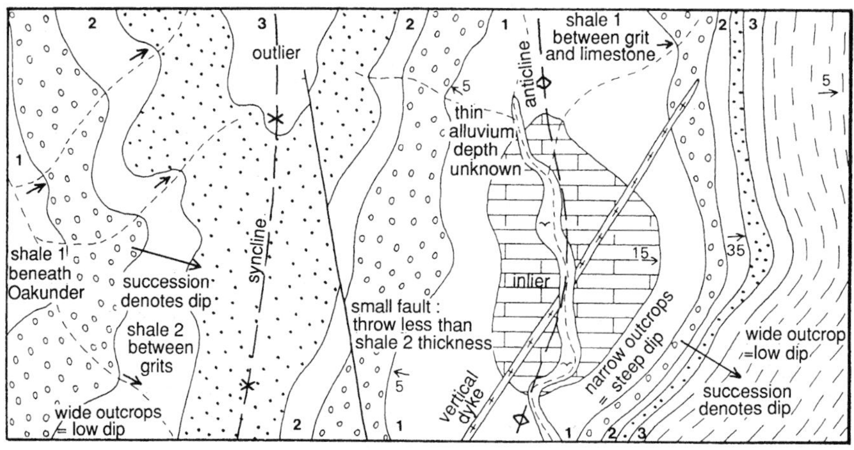

THREE-POINT INTERPRETATION

Sub-surface structure can be interpreted from a minimum of three isolated points (usually in boreholes) on a single horizon.
Assume locally uniform dip to draw stratum contours.
Along lines drawn between any pair of boreholes, distribute uniformly altitudes of the selected boundary or bed. Lines joining these interpreted points of equal altitude are therefore stratum contours.
With three boreholes, can only interpret and draw straight stratum contours.
With more than three boreholes, can draw curved and converging stratum contours to show folding and non-uniform dips. Reliability depends on borehole spacing in relation to structural complexity.
Useful for initial interpretation of site investigation data.
Applicable to any rock or drift layer, rockhead or fault.

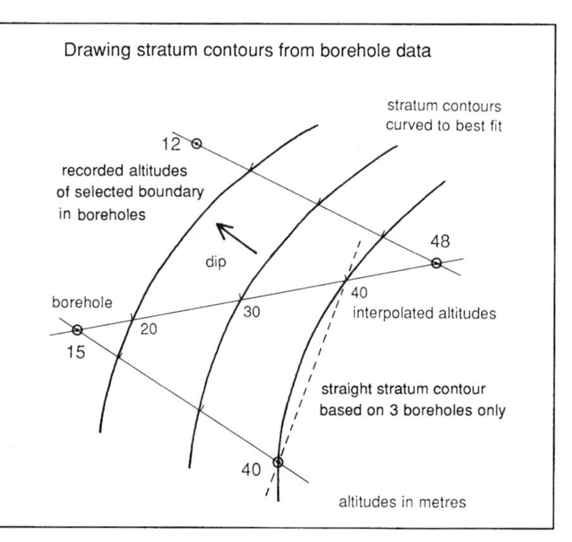

DRAWING THE SECTION

Sequence of stages for the Oakunder map is as follows:

1. Ground profile is given (or is drawn from topographic map).
2. Outcrops are projected onto ground surface (as in section 07).
3. Dips are obtained by using the given bed thicknesses (as on right).
4. Fault can be ignored at first because it is interpreted as minor.
5. Oakunder Grit is drawn across both folds, linking the three outcrops.
6. Fault is added so base of Belvoir Grit does not outcrop.
7. Oakunder Grit is adjusted across the fault within the syncline.
8. Other beds are added parallel, with constant thicknesses.
9. Parallel curves are fitted to changing dips east of anticline.
10. Slate marked below limestone thickness given in stratigraphic column.
11. Anticline core is unknown beneath given slate thickness.
12. Dolerite is added as vertical dyke cutting through beds.
13. Alluvium is given sensible thickness in valley floor.
14. Optional broken lines in the sky clarify structure.

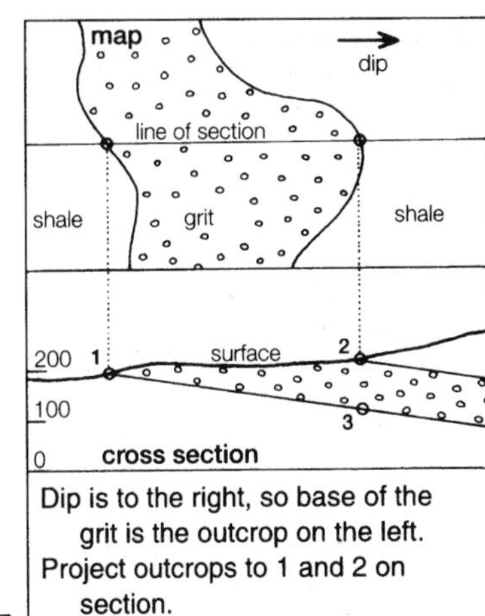

Dip is to the right, so base of the grit is the outcrop on the left.
Project outcrops to 1 and 2 on section.
Take thickness from the key to plot the base at 3, below the top at 2.
Draw the base of the grit through 1 and 3.

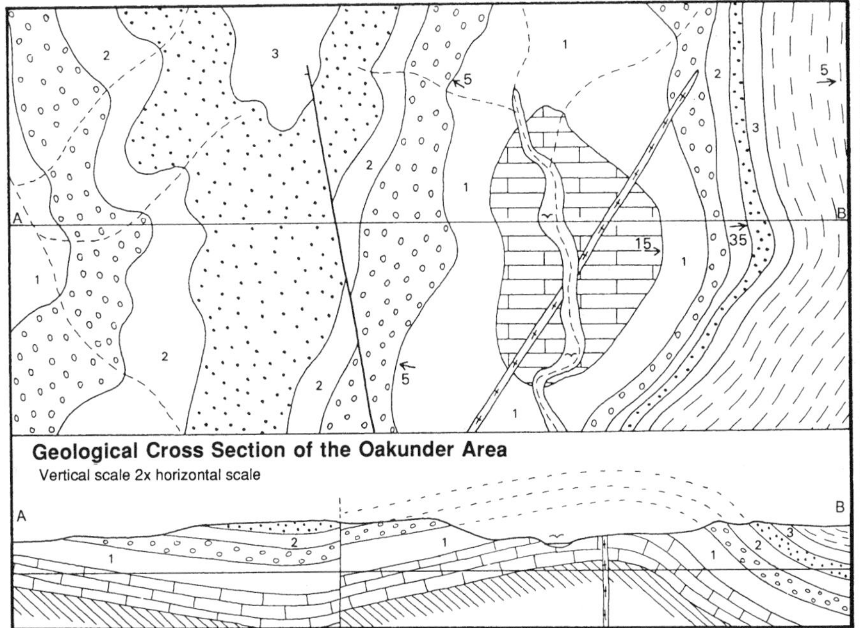

INTERPRETATION OF A STRATIGRAPHIC COLUMN

Data is symbolized to cover variations across the area on the map.
Drift deposits are at the top; their thicknesses are not shown.
Main rock sequence is drawn to scale to show mean thicknesses.
Sandstone E lies unconformably on the Carboniferous rocks.
Middle Coal Measures include sandstones and coals marked individually.
Undifferentiated MCM is a mixture of shales, mudstones and thin sandstones (this lithological data is only obtained from an accompanying handbook).
Sandstone D is locally absent where it is cut out by the unconformity.
Sandstone C varies from 6 to 16 m thick but is always present.
Sandstone B locally splits into two, and the upper unit may thin out.
Sandstone A varies from 0 to 12 m thick, and is missing in part of the map area.
Coal 2 locally splits, but the upper leaf may be cut out by sandstone.
Coal 1 has two leafs, with 5 m of shale between, over most of the map area.
Base of MCM is not seen on the map, an unknown thickness continues down.
Igneous rocks are at the bottom, even though they intrude into higher rocks.

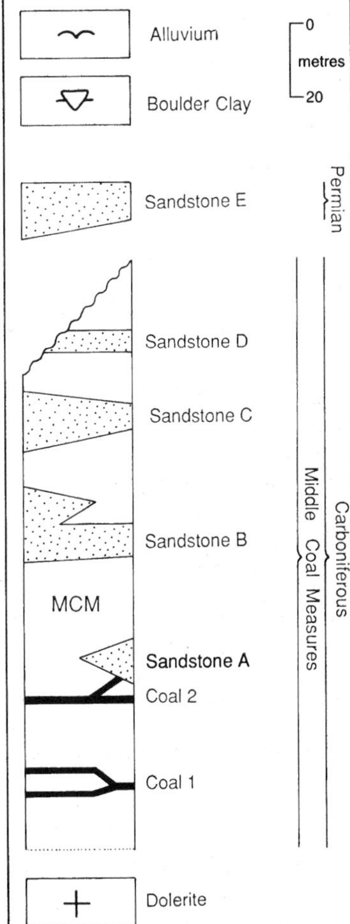

STANDARD GEOLOGICAL MAPS

Maps primarily record data observed at outcrops (or in boreholes). Interpretation is only added where needed; doubt is shown by broken lines. It is largely left to the reader to interpret the outcrop data.
Drift boundaries (see section 13) are distinguished from rock boundaries. Rock outcrops beneath drift may lack their map colour or ornament, but are labelled so that their boundaries are unambiguous.
Major areas of landslide and camber may be identified.
Larger scale maps may include underground data, borehole records, drift thickness and mine shafts etc.

ENGINEERING GEOLOGY MAPS

Extra data is available for some locations, with multiple map sheets covering the same area and individually showing selected features; these may include major rock properties, slope stability, subsidence potential, mine workings, drift thickness, drift bearing capacity, rock resources and groundwater conditions.
Generalized engineering geology maps may be summary compilations; these show secondary, interpreted, data to give useful broad pictures of ground conditions, but serve only as guidelines towards specific engineering site investigations.

09 Plate Tectonics

Planet Earth consists of three concentric layers:
Outer crust, < 100 km thick, various solid silicate rocks;
Mantle, 2800 km thick, hot plastic iron silicates;
Inner core, 3500 km radius, largely of molten iron.

CONTINENTS AND OCEANS

Oceanic crust is mainly basalt and dolerite, 5–10 km thick, forming all the ocean floors; it is created and destroyed at plate boundaries.

Continental crust is mainly granite and gneiss, 20–80 km thick; of lower density than oceanic crust, it floats higher on the mantle and forms all the continents, submerged continental shelves and adjacent islands. It is too light to be subducted, so is almost indestructible; it may be eroded or added to by accretion of sediment and rock scraped off subducting oceanic plates.

Individual plates may be either or both crustal types.
Continent coasts may or may not be plate boundaries.

THE MOHO

The boundary between the crust and the mantle is known as the Mohorovicic Discontinuity (or the Moho) recognized by refraction of seismic waves. No one has yet seen the rock beneath it. The American drilling project, the Mohole, was abandoned before reaching it, and the Russian borehole, 12·3 km deep by 1993, had not yet reached the Moho.

PLATES AND PLATE MOVEMENT

Lithosphere is the relatively brittle outer rock layer, consisting of the crust and upper mantle; it is broken into large slabs known as plates.

Convection currents circulate within the mantle – because it is heated from below – and the convection cells have horizontal movements over their tops.

Plate tectonics are the relative movements of the plates as they are shifted by the underlying mantle flows.

Each plate is relatively stable, but disturbances along the plate boundaries cause most geological processes. The formation of igneous, sedimentary and metamorphic rocks, and their subsequent deformation or erosion, can be identified on the cross-section diagram through two plate boundaries.

PLATE BOUNDARY TYPES

Conservative boundary has sideways movement only, e.g. San Andreas Fault. Major tear faults are formed, and intermittent movements create major earthquakes.

Divergent boundary is constructive, as new oceanic plate is formed, e.g. Mid Atlantic Ridge. Basaltic magma is produced as a silicate liquid separated from iron rich mineral solids in partially melted mantle; this produces numerous dykes and submarine volcanoes. Excess magma creates islands, e.g. Iceland, with effusive basalt volcanoes and high geothermal head; small earthquakes occur as rocks part under tension.

Convergent boundary is destructive, as oceanic plate is subducted and melted, e.g. beneath the Andes along western edge of South America. The over-riding continental plate is crumpled and thickened to form a mountain chain, involving a great range of geological processes, collectively known as orogenesis (from the Greek for mountain building).

Type of convergence determines the style of orogenesis:
Continent–ocean: normal orogenic belt, ocean destroyed, e.g. Andes.
Ocean–ocean: one plate destroyed, magma creates island arc volcanoes, e.g. Java.
Continent–continent: collision, orogenic maximum, welds plates together, e.g. Himalayas.

MOUNTAIN CHAINS

Uplift of mountain chains occurs because the lightweight granitic crust, thickened within the orogenic belt, flows to a higher level on the mantle in order to maintain the isostatic balance of equal loading all around the rotating sphere of Earth.

The highest mountain chains are the youngest. The Himalayas are < 10M years old, formed largely of folded sedimentary rocks; top of Mount Everest is limestone.

Old mountain chains are eroded down. The Scottish Highlands have been eroded for 400M years; they consist of granites and gneisses, rather similar to the rocks on the floors of the deepest Himalayan valleys.

The world's major crustal plate, with arrows to show relative movement (mostly a few cm/year)

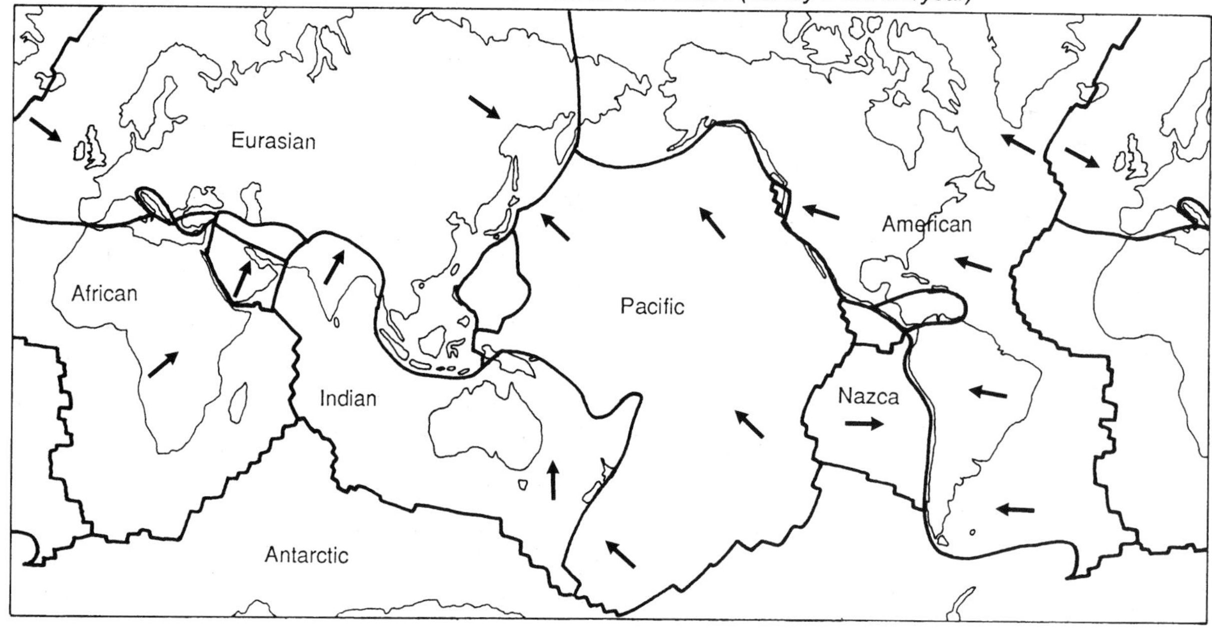

GEOLOGICAL ENVIRONMENTS

The overall geological character of a region – whether or not it has metamorphic rocks, active volcanoes or earth movements, whether the sedimentary rocks are thick or thin, folded or unfolded – relates to the plate tectonic processes.
These are the background to the ground conditions of concern to the engineer.
Stable environments are on the plates; the sedimentary rocks, slow erosion, gentle folding, only rare earthquakes and probably no volcanoes. The oldest continental plates are the shields of strong, basement, metamorphic rocks forming Scotland's Outer Hebrides and most of eastern Canada. Britain and the eastern USA are on younger stable plates.
Unstable environments are on or near plate boundaries, and the geology relates to the boundary movement: sideways, divergent or convergent.

OROGENESIS

Involves all the main geological processes except basaltic volcanoes.
Strong folding, overfolds and nappes; weaker folds away from boundary.
Faults and thrusts under compression, and major earthquakes.
Regional metamorphism by heat and pressure at depth.
Partial melting of continental crust creates granite batholiths in cores of metamorphic belts.
Melted oceanic basalt mixes with continental material to form viscous andesite and rhyolite magmas and explosive volcanoes.
Mineralization by migrating fluids in hot active zones.
Uplift of mountain chain; consequent rapid erosion and sediment production.
Thick sedimentation in adjacent subsidence zones; turbidites into oceans.

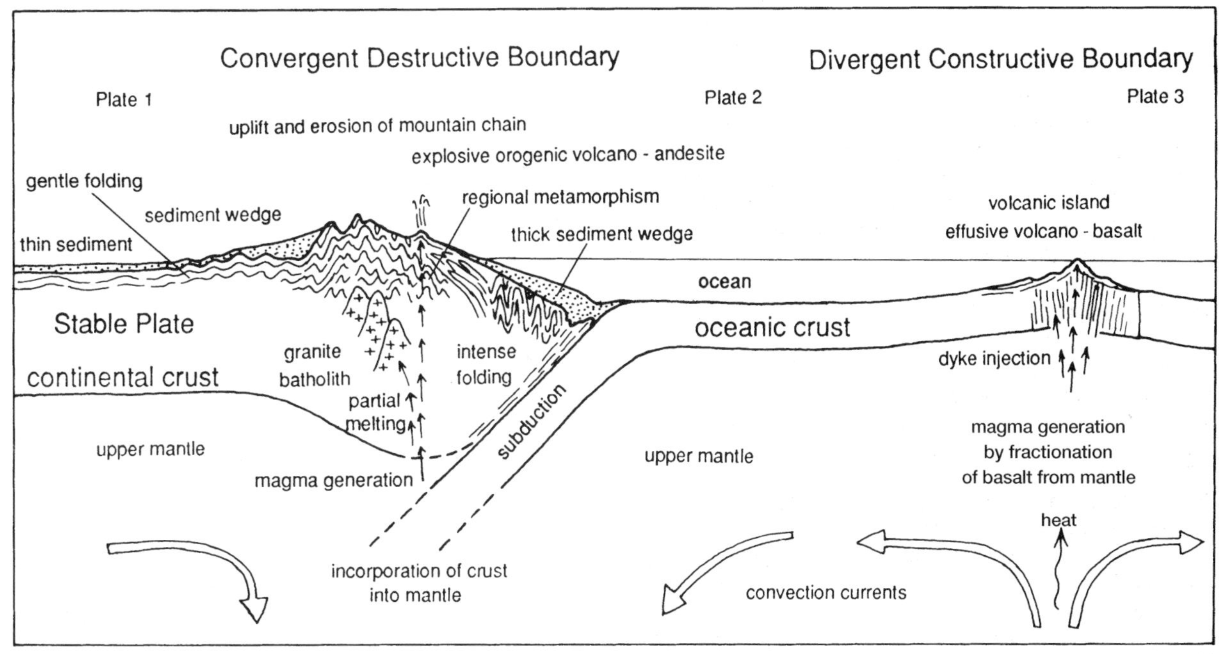

Plate boundary processes

GEOLOGICAL TIME

Processes evolve and plate patterns change over geological time. New oceans open up; continents collide and weld into one. Any one place can therefore be in a series of geological environments through its history.
Time is an extra dimension which must be appreciated to fully understand the geology of any area; the geological history of a site accounts for the structures and rock relationships, which are relevant to ground engineering.
The significance of rock age to an engineer: in any one area, e.g. Britain, older rocks are generally stronger, better lithified, maybe metamorphosed, and more complexly folded than younger rocks. However, rocks of the same age may be very different in areas of different plate tectonic histories, e.g. the contrasts between the rocks of the east and west coasts of the USA.
Quaternary sediments are so young that most have not been deeply buried; they are largely unconsolidated and minimally deformed.
The stratigraphic column divides geological time into periods, and the same names apply to the systems of rocks formed in those periods.
The names are international, except that Carboniferous is replaced by Mississippian and Pennsylvanian in the USA.

THE STRATIGRAPHIC COLUMN

period/system	era	age M.y.
Holocene = Recent Pleistocene	Quaternary	
		1·8
Pliocene } Neogene Miocene Oligocene } Paleogene Eocene Paleocene	Tertiary	
		65
Cretaceous Jurassic Triassic	Mesozoic	
		245
Permian Carboniferous { Pennsylvanian, Mississippian Devonian Silurian Ordovician Cambrian	Paleozoic	
		545
Proterozoic Archean (origin of the earth)	Precambrian	4600

10 Boundary Hazards

Earthquakes

Caused when relative movement of plates or fault blocks overcomes shear resistance of a fault. Movement builds up elastic strain in rocks; fault rupture and rock rebound release strain energy as ground shock waves.
Most earthquakes originate at focus < 20 km deep.
Surface displacement may be a few metres or absent.
Fault breaks may extend over lengths 1–100 km.

SIZE AND SCALE OF EARTHQUAKES

Ground movement is measured in different planes on seismographs.
Magnitude defines the size of an earthquake on the Richter scale: $\log_{10}$ of the maximum wave amplitude in microns on a Wood Anderson seismograph 100 km from the epicentre (point on the surface above the focus).
Moment magnitude relates to fault area, movement and rock rigidity – a better indication of earthquake's energy.
Intensity is the scale of earthquake damage at any one point, described on the modified Mercalli scale, and declining away from the epicentre.
MSK intensity is similar to Mercalli, but with more detail.
Damage relates largely to peak ground acceleration, also to peak velocity, frequency and duration.
Duration usually < 10 s for magnitude 5, may last 40 s for magnitude 8; increases away from epicentre.

Mercalli Earthquake Intensity (and peak acceleration)

I	Not felt	VII	Adobe damaged (~0.1g)
II	Felt at rest	VIII	Masonry damaged
III	Felt indoors	IX	Foundations damaged
IV	Windows rattle (<0.02g)	X	Buildings destroyed (>0.6g)
V	Felt outdoors	XI	Railways buckled
VI	Frightening	XII	Total destruction

Types of Earthquake Waves

P and S are small, fast interior waves, from which form Love and Rayleigh, which are larger, slower, and more destructive surface waves.
Velocity difference of P and S creates time lag on a siesmograph, so distance to epicentre is calculated at about 9 km/s of the lag.

EARTHQUAKE PREDICTION

Most are on plate boundaries; 90% on subduction zones.
Some occur on intraplate faults: Britain has up to M5, and the Mississippi Valley earthquakes of 1811 reached M7·8.
Also due to magma movement under volcanoes.
Some faults slip smoothly: the Cienega Winery in California has its foundations displaced 15 mm/year by the San Andreas Fault – but no earthquake damage.
Tangshan quake, China, 1976, killed > 250 000 people.
Prediction: research is now greatly reduced, in favour of research into structural survivability in earthquakes.
Side effects of ground strain before some quakes may include foreshocks, uplift, dilation, gas emissions, groundwater changes and increase in seismic velocity, but monitoring reveals inconsistent patterns.
Historical data may indicate seismic gaps (with no recent movement) on an active fault, where a future earthquake is more likely.
Control: raised water pressures reduce shear strength, and cause fault movement before large strain energy accumulation. Pumping water into deep wells does trigger premature small quakes, but legal complications make serious earthquake control impossible.

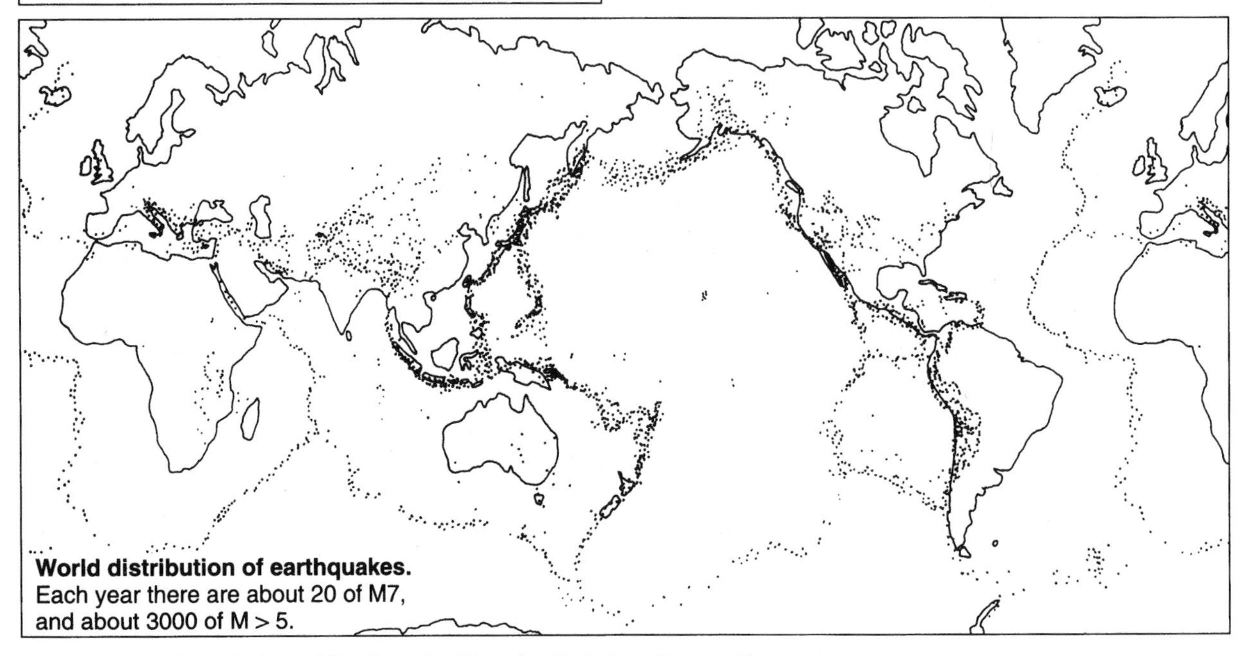

World distribution of earthquakes.
Each year there are about 20 of M7, and about 3000 of M > 5.

Approximate Correlation of Earthquake Magnitude, Intensity and Damage

Relative energy	Richter Magnitude	Example	Maximum Intensity	Damage at epicentre	Area of influence
	1–3	–	I – IV	Social disturbance, no damage	Limited
1	4	1979 Carlisle M4·8	VI	Slight	Limited (Major earthquake for Britain)
30	5	1979 San Francisco M5·9	VII	Little damage to reinforced concrete	(Minor earthquake for California)
				Severe damage to adobe houses	Intensity VI slight damage to 10 km away
1000	6	1971 San Fernando M6·5	IX	Severe damage to many buildings	Intensity VII damage to 10 km away
30 000	7	1970 Chimbote (Peru) M7·7	X	Major damage to most buildings	Intensity VII damage to 50 km away
1 000 000	8	1906 San Francisco M8·3	XII	Total destruction	Intensity VII damage to 200 km away
					Intensity X severe damage to 20 km away

CONSTRUCTION IN SEISMIC ZONES

Adobe and dry stone walls fail under horizontal acceleration of 0·1g, but good low rise timber buildings can withstand any earthquake.
Reinforced concrete structures need bracing to stop rhombohedral collapse; this can be provided by massive, resistant shear walls, or diagonal steelwork.
Rebars must be integrated across intersections of columns/beams/walls/slabs.
Pile cap failures are restrained by tie beams and integrated basement structures.
Buildings and bridges can be isolated on rubber spring blocks; and steel springs can act as energy absorbers to stabilize structures.
Precautionary provisions add 5–10% to construction costs. Later modifications are more expensive.
Avoid ornamental appendages which can fall off.
Use land zoning to avoid areas of deep soft soils and known fault traces – any displacement of Holocene soils indicates modern activity on a fault.
New building in California is prohibited within 15 m of active faults; wider zones apply to larger buildings and less well-mapped faults.

DEEP SOILS AND EARTHQUAKES

Soft soils do not dampen ground vibrations. They amplify them. Buildings on soft soil suffer much worse earthquake damage than those on bedrock.
Wave amplitude may double passing from rock to soil.
Dominant natural period of the shock waves also increases, from about 0·3 seconds in solid rock, to 1–4 seconds on soils. The natural period further increases with soil depth, and with distance from the epicentre.
Buildings have a natural period of about N/10 seconds (N = number of stories). Maximum damage is due to resonance, when periods of building and soil match.
Deep soft soils have long periods which match those of high rise buildings susceptible to more catastrophic damage – as in the Mexico City earthquake in 1985.
Compared to adjacent bedrock, soft soils cause damage 1–3 intensities higher.

Secondary earthquake phenomena

Subsidence due to liquefaction of low density sands.
Landslides and slope failures of all sizes and speeds.
Tsunamis – oceanic seismic waves (section 17).
Seiches – oscillating waves on lakes.

Volcanic Eruptions

Basaltic volcanoes lie on divergent plate boundaries (e.g. Iceland), or on plates away from boundary disturbances (e.g. Hawaii), where magma is generated from mantle plumes. They produce large flows of mobile lava in quiet, effusive eruptions, with only limited fountaining or explosions.
These volcanoes are tourist attractions, which may threaten fixed structures, but offer minimal threat to life.

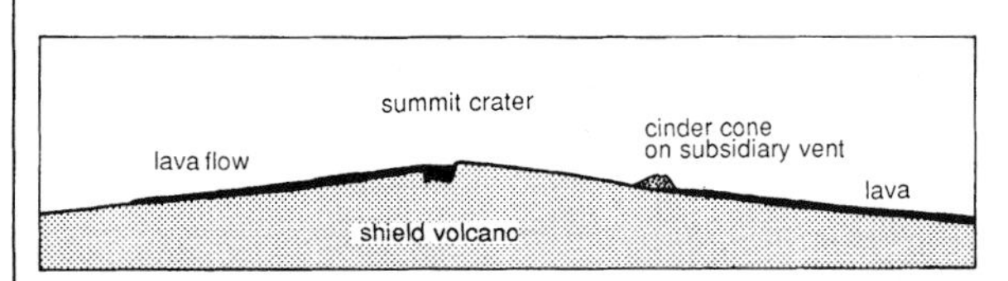

Prediction of eruptions is largely based on volcanic inflation (uplift) and seismic monitoring, with successful forecasts of repetitive basalt emissions. Scale and size of explosive eruptions cannot be reliably predicted, nor can their precise timing and location within the volcanic area.

Explosive volcanoes all lie on the convergent plate boundaries (*eg* Krakatoa, St Helens), where magma is generated by subduction melting. Viscous magma, of andesite or rhyolite, makes gas pressures build up.
Eruptions produce high ash clouds, explosive blasts and very dangerous pyroclastic flows (of hot gas and ash) which turn into lahars (mud flows) lower down valleys; lava flows are minor and short.
Flank collapses can cause massive lateral blasts.
These eruptions are dangerous, largely unpredictable and totally uncontrollable; they must be avoided.

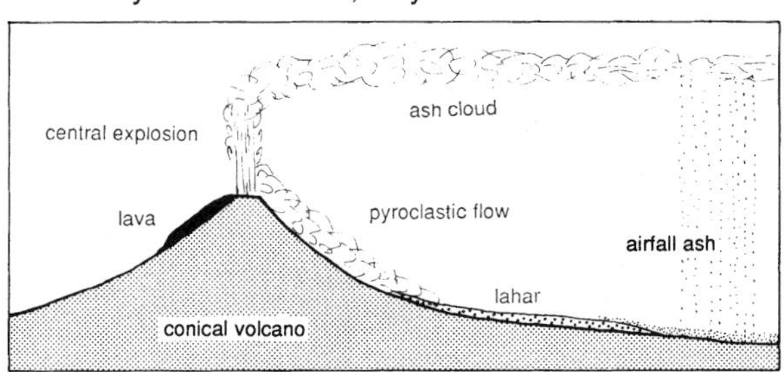

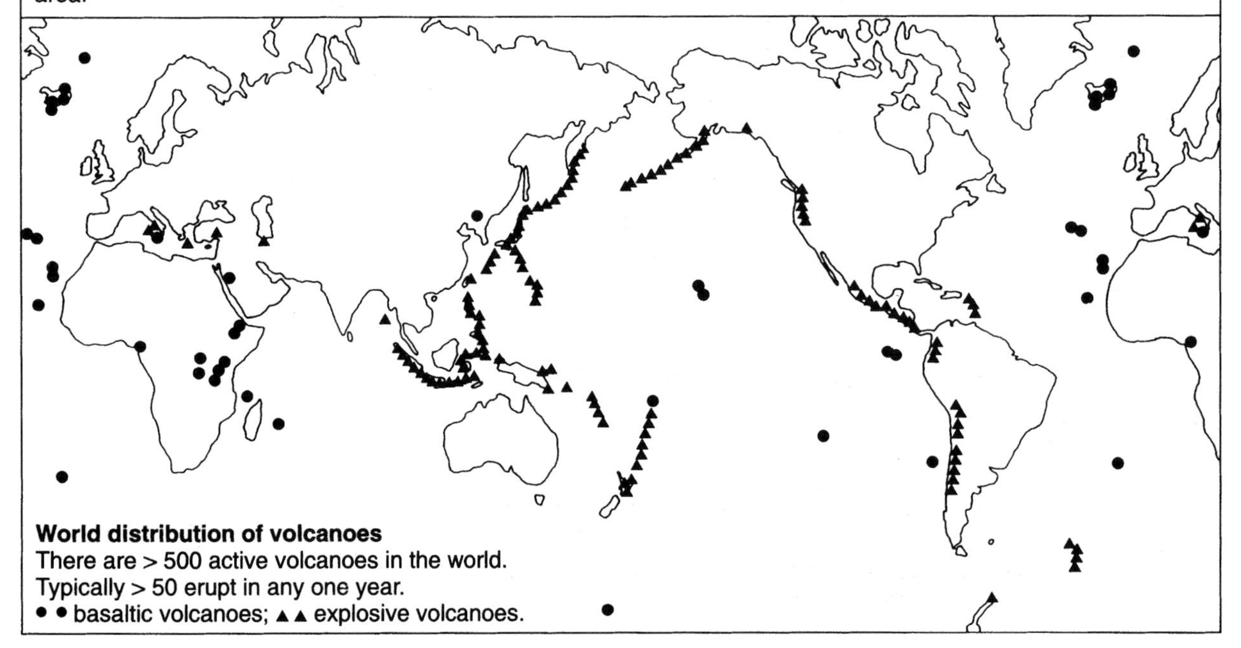

World distribution of volcanoes
There are > 500 active volcanoes in the world.
Typically > 50 erupt in any one year.
● ● basaltic volcanoes; ▲▲ explosive volcanoes.

11 Rocks of Britain

Britain covers an area small enough to have its geology viewed as a single sequence of processes, encompassing the whole country. With a single history, the geology of Britain is sensibly divisible by rock age, and with few exceptions the older rocks are stronger and more deformed than the younger. For such a small area there is amazing diversity within the geology, and all ages are represented within the rocks.

The tectonic framework of Britain has evolved over two successive convergent plate boundaries and then a divergent boundary; these have annealed fragments of continental crust to create the complexity of Europe, followed by the western breakaway of the Atlantic opening and the tensile thinning of the North Sea crust. This evolution has created major contrasts across the country:

The old rocks of the northwest:

- huge thicknesses of rock crumpled on convergent plate boundary 400 million years old;
  have formed land subject to erosion ever since;
- now strong metamorphic rocks, intensely folded;
- accept high bearing pressures;
  yield valuable stone and aggregate resources.

The young rocks of the southeast:

- thin sediment sequences formed on the edge of subsiding North Sea basin less than 200 million years old;
  mostly covered by sea until 25 million years ago.
- now weak sedimentary rocks, gently folded;
- can take only low foundation loading;
  have no good aggregate resources.

The Carboniferous rocks of the middle:

- thick sediment sequences formed on wedge of plate between two boundary disturbance zones;
  include the Coal Measures of Britain's industrial heartland;
- now strong sedimentary rocks, well folded;
- very varied ground conditions;
  yield valuable rock resources of all types.

Geological evolution of Britain can be seen in time sequence of changing patterns of plate boundary processes and sedimentary environments.

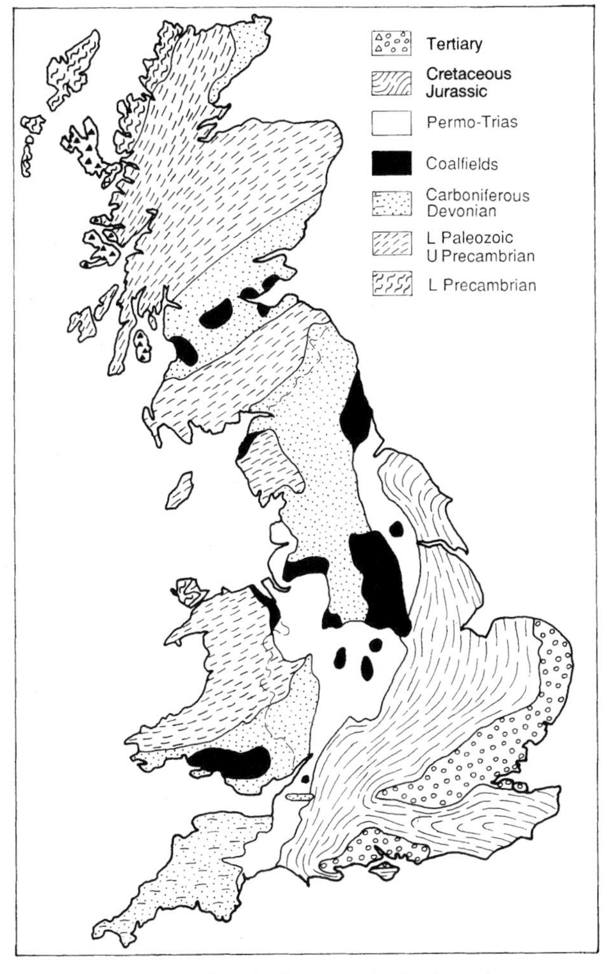

The map divides Britain into geological environments, largely related to age but primarily distinguished by the rock types and structures, which are the main concern of the ground engineer. The marked coalfields include concealed parts beneath Permian and Triassic cover.

Ireland represents a western continuation of the geology of Scotland and northern England; it is dominated by Carboniferous and older rocks, with the Antrim basalt plateau covering them in the north.

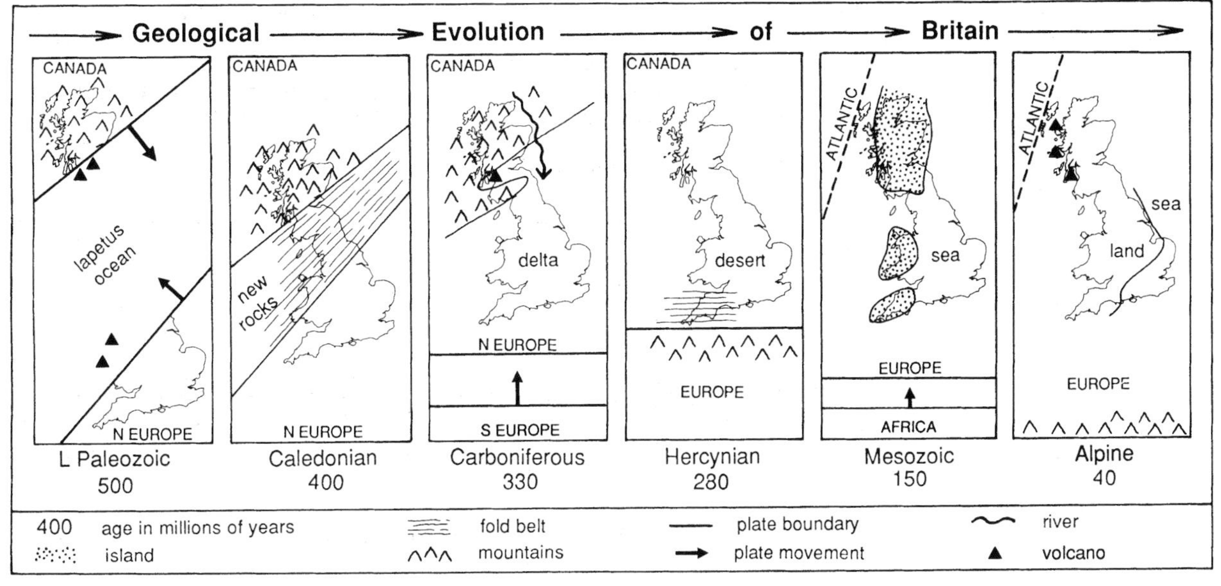

MAJOR ROCK UNITS OF BRITAIN	**Tectonics and environment**
QUATERNARY Unconsolidated sand and clay, alluvium and till.	Coastline as now; Ice Ages.
TERTIARY Soft sediments of London and Hampshire basins. Poorly consolidated sands and clays, with 200 m thick London Clay. Also basalt lavas, volcanic centres and intrusives of western Scotland.	Gentle Alpine folding of all rocks. Deltaic sediments in bays temporarily flooded by North Sea. Volcanoes on Atlantic divergence.
JURASSIC and CRETACEOUS Weak sedimentary rocks forming most of southern and eastern England. Chalk – 200 m thick, soft limestone with flint horizons forming Downs, Chilterns. Clays and sandstones of Weald and Midlands, with thick Oxford Clay beneath Fenlands, unstable Gault Clay, and Portland limestones in south. Sandstones and ironstones of Midlands and North Yorkshire Moors. Oolitic limestones (oolites) from Cotswolds to Lincoln ridge. Lias blue-grey clays with thin limestones in Midland lowlands.	Uplift forms land almost as today. Thin sediment accumulation in shallow seas over England. Sea forms shelf, marginal to subsiding North Sea basin Scotland and Wales form islands, with no deposition. Atlantic opening starts. Submergence under sea.
PERMIAN and TRIASSIC Red sandstones and mudstones of Midlands lowlands. Red mudstones with beds of salt and gypsum, including Mercia Mudstone. Yellow and red sandstones with conglomerates, including Sherwood Sandstone. Magnesian Limestone – impure, sandy or dolomitic limestone east of Pennines. Granites of Devon and Cornwall with associated mineralization.	Desert sediment accumulation: salt playas in low relief; alluvial fans around mountains. Marine incursion from east. Magma from orogenic core.
CARBONIFEROUS Strong sedimentary rocks forming most of the high ground of Northern England, South Wales and Central Scotland, including all coalfield industrial areas. Clyde Valley basalt lavas, Edinburgh volcanics, Whin Sill dolerite of Pennines. Coal Measures – 2000 m of repetitive cyclic sequences, of sandstones, siltstones, mudstones, dark shales and thin coal seams (up to 2 m thick). Millstone Grit Series – alternating sandstones (grits), shales with flagstones. Limestone - massive limestones with cherts in S. Pennines, Wales and Mendip; thin bedded impure limestones and shales in N. England and Scotland. Slates and grits in Cornwall and Devon.	Hercynian folding includes Pennine anticline and coal basins; more intense with metamorphism towards plate boundary in south Marginal plate boundary disturbance. Intermittent swamp forests established on subsiding delta flats. Massive delta expanding from north. Shallow shelf seas and basins. Convergent plate boundary in south Sediments in subducting ocean zone.
DEVONIAN Slates, grits and limestones in Cornwall and Devon. Brown sandstones and basalt lavas of Tayside and Ochils. Red and brown sandstones and mudstones of Brecon Beacons and Orkney. Highland granites west of Aberdeen.	Marine sediments in southern ocean. Red beds in desert basins surrounded by new mountains. Melting in orogenic core.
LOWER PALEOZOIC Mountains of Wales, Lake district and Southern Uplands; repeated greywackes and slates 12 km thick; include Ffestiniog slate, Wenlock limestone, rhyolite and andesite lavas and tuffs of Snowdon and Borrowdale Volcanics. Dalradian schists, gneisses and marbles in southern part of Scottish Highlands.	Strong metamorphism of all old rocks. Caledonian folding at plate collision. Subduction of Iapetus Ocean plate, beneath two convergent boundaries, with local belts of volcanics. Northwestern boundary more active.
PRECAMBRIAN Moine schists and gneisses forming most of Scottish Highlands. Torridonian Sandstone of northwest Scotland. Buried basement of England and Wales, only exposed in small inliers, including Charnwood, Birmingham, Long Mynd and Anglesey. Lewisian basement gneisses of northwest Scotland and Outer Hebrides.	Active convergent plate boundaries. Sands deposited on continental block. Old continental blocks: S.E. block now largely buried; fragment of N.W. block exposed.

To follow a time sequence, this table should be read from bottom to top.
Sequence of rocks in stratigraphic order, with youngest at top.

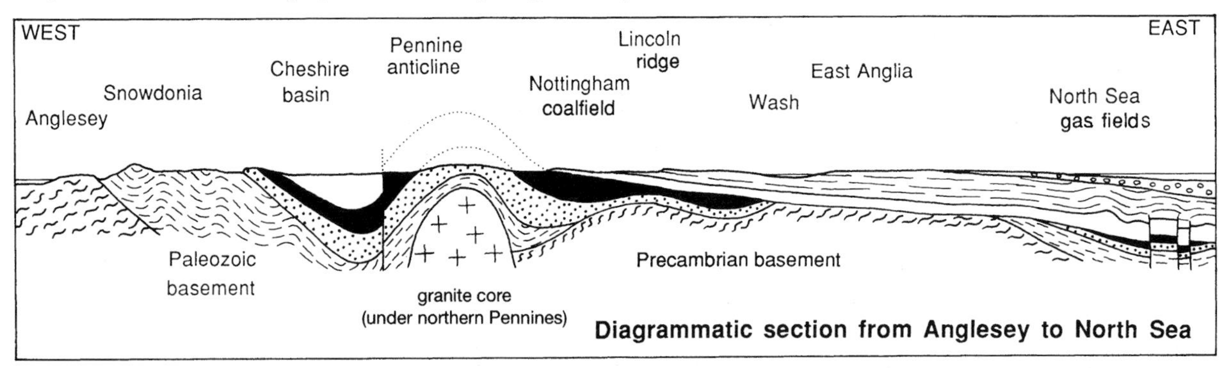

Diagrammatic section from Anglesey to North Sea

12 Rocks of the United States

The USA spans an entire continent and every type of geological environment. Unlike Britain its geology cannot be viewed in a simple time sequence: at any one time plate boundary activity on one side could be distant enough to leave the other side unaffected. Over such a large area, age alone is meaningless – igneous and metamorphic rocks in the west are far younger than barely folded sedimentary rocks in the east.

It is sensible to divide the USA into geological provinces, each with its own character and geological history, distinguished from its neighbours by both the types and structures of its rocks.

The North American continent has a core of strong Precambrian basement rocks: these form the Laurentian shield of Canada, and underlie the Great Plains and Central Lowlands where thin sedimentary covers were deposited when shallow seas lapped onto the basement core.
Thick sedimentary rock sequences were crumpled and metamorphosed along plate boundaries and welded onto both sides of the core: the older Appalachia on the east, and the younger Rockies and Cordillera on the west. The active western plate boundary has been complicated by the continental slab obliquely overriding the rising convection plume of the East Pacific rise, creating the tension in the basin and range province and adding to the volcanic activity.

COAST RANGES
Complex series of deformed rocks along the active plate boundary of the west coast. In California the highly folded Franciscan greywackes and schists include a slice adjacent to the Sierra Nevada containing the gold mineralization of the 1849 rush.
Faults include the very active San Andreas zone with its associated earthquakes and break the ranges into fault blocks further south. In the Los Angeles area, basins have up to 5 km of oil-bearing Tertiary sediments between older mountain blocks. The Peninsular Ranges have granite intruded into the metamorphic rocks.
Further north, the Klamath Mountains are fault slices of Paleozoic sedimentary, metamorphic and igneous rocks, followed by Tertiary basalts along the Oregon coast, and Tertiary granites and schists forming the Olympic Mountains.

CENTRAL VALLEY OF CALIFORNIA
Faulted basin containing 12 km of Mesozoic sedimentary rocks. Surface has up to 400 m of poorly consolidated Quaternary sands, silts and clays causing widespread subsidence problems.

CASCADES
Chain of Tertiary and modern volcanoes above the active subduction zone of the convergent plate boundary. They include Lassen and St Helens with major explosive eruptions in the last 100 years. Mainly andesite and rhyolite lavas, with locally thick pyroclastics and some basalts.

COLUMBIA PLATEAU
Volcanic province dominated by horizontal Tertiary flood basalts. Eastern extension has younger Snake River Pleistocene basalts, reaching to more varied pyroclastics and lavas of Yellowstone Park with its continuing geothermal activity.

IDAHO BATHOLITH
Strong, massive, Cretaceous granite forming the largest batholith in the Rockies.

SIERRA NEVADA
Cretaceous batholith of massive, strong granite, with glacial features on much of the high ground.

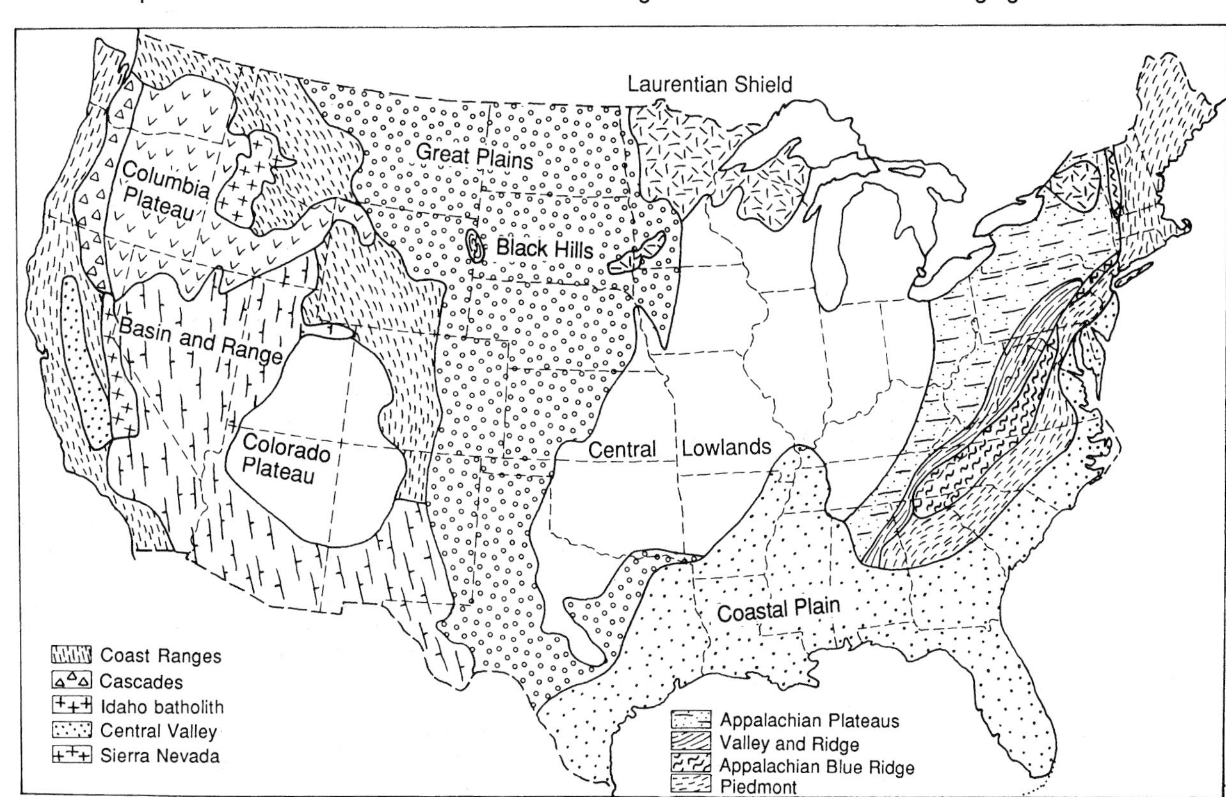

ALASKA
Essentially an extension of all the units and provinces in the Western Cordillera. Interior mountain chains of folded Paleozoic rocks, older metamorphics, and younger intrusions separate basins with thick sedimentary sequences. Coast ranges consist of younger metamorphics along with active volcanoes and faults.

HAWAII
Islands are tops of huge ocean floor volcanoes formed in a chain where Pacific plate is moving over a mantle hot spot. Rocks are nearly all basalt, with minor tuffs, alluvium and reef limestones; active volcanoes are now only on Hawaii Island.

LAURENTIAN SHIELD
Southern tip of the exposed Precambrian basement which forms much of eastern Canada. Complex of strong, deformed rocks, dominated by greenstones, slates, greywackes and granites, with an extensive cover of Pleistocene till. Includes an exposed inlier in South Dakota and the Adirondaks of New York.

GREAT PLAINS
Flat lying Mesozoic and Tertiary rocks on a Precambrian basement. Mainly shales and weak sandstones, with some coal basins and localized volcanics. Include the clays of the Badlands and the Cretaceous limestones of the Edwards Plateau. Pleistocene sands, clays and loess in the south and centre are replaced by glacial till in the north.

APPALACHIAN PLATEAUS
Strong sandstones, shales and limestones, mostly of Upper Paleozoic age. Gently folded with low to moderate dips. Valuable coal seams in the Pennsylvanian rocks are extensively mined in all of the province except New York state.

VALLEY AND RIDGE
Strongly folded Paleozoic rocks in a sequence 18 km thick. Long, steep, escarpment ridges of strong sandstone and limestone separate parallel valleys in softer shales and coal measure rocks.

APPALACHIAN BLUE RIDGE
High grade metamorphic Precambrian basement rocks overthrust onto younger rocks to the west. Mainly gneisses, but with thick greenstones in Shenandoah Mountains and strong metamorphosed sandstones in the Great Smokies. Includes extensions of gneisses north of New York and into Vermont.

PIEDMONT
Eastern flanks and foot slopes of the Appalachians, formed mostly of strongly deformed Paleozoic slates and greywackes, with igneous intrusions in the Carolinas. More highly metamorphosed rocks include the Manhattan gneiss and more extensive schist belts. Province includes New England with less metamorphosed rocks and a number of granites. Pleistocene sediments and glacial tills form Cape Cod and much of Long Island.

BLACK HILLS
Anticlinal inlier with Mississippian limestones around a core of schists intruded by the granite of Mt Rushmore.

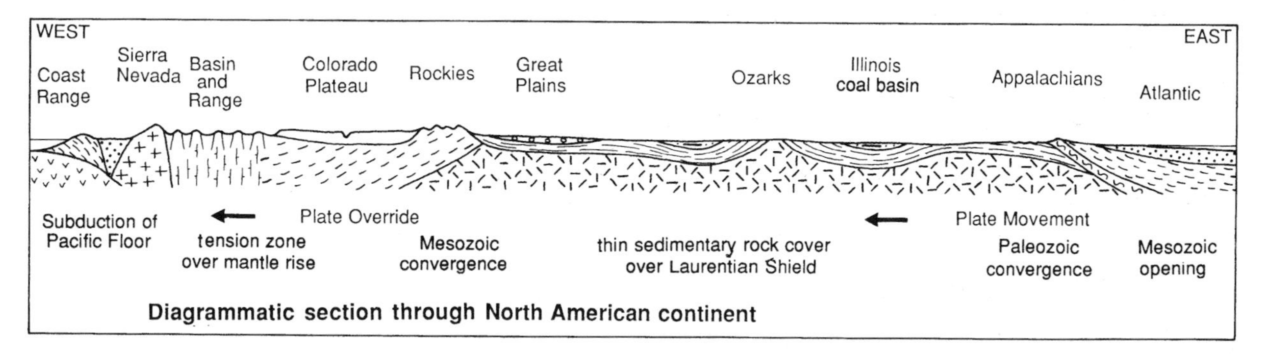

Diagrammatic section through North American continent

ROCKY MOUNTAINS
Precambrian and Paleozoic metamorphics forming the old core of the Cordillera. Generally strong rocks include gneiss, schist, greenstone and marble, mostly well folded, with various younger intrusives and volcanics. Basins in Wyoming contain thick Tertiary sedimentary rocks with coal and oil shale.

BASIN AND RANGE
Vast province of fault blocks broken by east-west tension within the crust. Upfaulted ranges of strong igneous and metamorphic rocks of Precambrian to Mesozoic age, and also the Permian limestones of the Guadalupe Mountains. Downfaulted basins contain weaker Tertiary and Quaternary clastic sediments, along with the salt sequences of the Great Basin, Death Valley and elsewhere, and some volcanics along fault zones.

COLORADO PLATEAU
Flat lying, Carboniferous to Tertiary, sedimentary rocks, overlying basement schists exposed in the Grand Canyon. Includes the strong red sandstones of Canyonlands and national parks, soft limestones at Bryce Canyon, scattered basalt volcanics, and some shallow coal basins.

CENTRAL LOWLAND
Paleozoic sedimentary rocks gently folded into broad basins and domes. Limestones of the Ozarks form the largest dome, with an exposed core of Precambrian gneiss, and also host extensive mineral deposits. Further east, strong Mississippian limestones form extensive plateaus. Pennsylvanian coal measures, dominated by shales, siltstones and strong sandstones, are up to 5 km thick in the main coal basins west of the Ozarks, across Illinois and in Michigan. Extensive Pleistocene cover of glacial till in the north, and loess in the southwest.

COASTAL PLAIN
Poorly consolidated Tertiary clastic sediments provide no hard rock resources. Widespread subsidence problems occur on extensive soft clays, on some large areas of peat, and over the Tertiary limestones forming much of Florida. Mesozoic sedimentary rocks outcrop along the Appalachian margin and thicken towards the coasts beneath a Tertiary cover. Up to 6 km of sediments underlie the Gulf Coast, containing salt domes and major oilfields, together with extensive lignite (brown coal) resources in Texas. Younger deltaic sands, clays and peats fill the Mississippi basin.

13 Weathering and Soils

GROUND CONDITIONS

Top few metres of the ground profile generally consist of soil, drift and weathered rock, with engineering properties very different from those of the underlying bedrock.

Soil: mixture of weathered mineral debris and plant material, usually <1 m thick; may divide into plant-rich topsoil and clay-rich subsoil.

Weathering: the natural decay and breakdown of rock or drift in contact with air and water; generally <10 m deep.

Drift: transported, superficial sediment deposited on top of the bedrock; mostly unconsolidated clay, sand and coarser clastic debris; generally Quaternary age, hence too young to be consolidated; varies in thickness from 0 to > 50 m.

Colluvium: slope debris, moved downslope largely by gravity alone; extent of sediment transport therefore drift > colluvium > soil; includes debris from creep and sheetwash, also head and scree. Sheetwash by surface water increases greatly with loss of vegetation.

Rockhead: the buried drift/rock interface; commonly a conspicuous boundary between weak soils and drift and strong rock; may be less well defined in deep profile of weathered rock; formed as erosion surface before drift deposition so its topography may be totally unrelated to modern surface.

Engineering soil: weak material (UCS < 600 kPa) that can be excavated without ripping or blasting, therefore including soil, drift, weak rocks and weathered rocks.

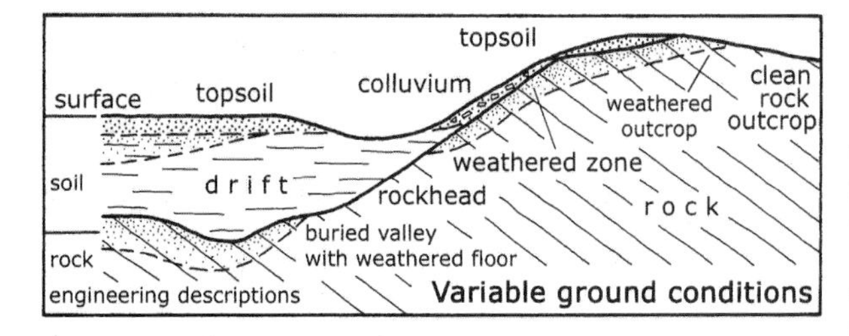

ROCK WEATHERING

Physical and chemical breakdown of rocks at or near the surface. Subsequent removal causes surface lowering:

Weathering + Transport = Erosion

Weathering processes depend on contact with air and/or water, so are strongly influenced by climate.

- Frost shattering is important in cooler latitudes and higher altitudes.
- Salt crystallization is only significant in deserts with high evaporation.
  All chemical processes accelerate in hot wet climates, and are further increased by organic acids from dense plant cover.
- The most important chemical process is the production of clay minerals from other silicates.
- Temperate weathering produces illite as the dominant clay mineral.
- Hot wet weathering of igneous rocks produces the unstable smectite.
- Laterite: red soil, high iron and aluminium, low silica, formed in tropics.
- Saprolite: totally decomposed rock retaining ghosts of original structure.
- Spheroidal weathering: forms rounded boulders or corestones from angular joint blocks weathered more at edges and corners.

DEPTH OF WEATHERING

Depends on the timescale, rock type and climate.

Rocks only exposed for 10 000 years (since last glaciation) are less deeply weathered than those exposed for a million years in unglaciated areas.

Shales, porous sandstones and weak limestones weather to greater depths than do granites and compact metamorphic rocks.

Deepest weathering occurs under climatic extremes, of either periglacial frost action or beneath equatorial rain forest.

Top of zone II is effectively rockhead, but is not sharply defined; it is usually about 1–5 m deep in Britain; but zone I fresh rock may only be found at depths > 20 m in quarries which demand the best quality of rock.

In tropical areas, soils of zone IV may reach depths of 5–20 m. Decomposed granite of weathering grade III commonly reaches > 30 m deep in Hong Kong.

This road cutting in Hawaii shows an almost complete weathering sequence in basalt lavas.
Grade III material is not seen in this sequence, because a change of rock type is more significant than the weathering state – a layer of weak, rubbly, scoriaceous lava has weathered much more completely than the solid lava above it.
Grade I fresh rock only occurs at greater depths, below this cut face. For engineering purposes, sound rock is found near the top of zone II, about 4m below the surface at this site.

Physical Weathering

Unloading joints:	stress relief fractures due to overburden removal.
Thermal expansion:	fracturing due to daily temperature changes.
Frost shatter:	fracturing as fissure water or porewater freezes and expands.
Wetting and drying:	movement due to loss or gain of water in clays.
Root action:	tree root expansion in fissures, and rootlet growth in pores.
Crystallization:	growth of salt crystals where groundwater evaporates.

Chemical Weathering

Solution:	mainly of calcite and gypsum, in sandstone cement, veins and limestone.
Leaching:	selective removal of solutes or specific elements.
Oxidation:	notably rusting and breakdown of iron.
Hydrolosis:	most silicates react with water to form clay minerals.

ENGINEERING CLASSIFICATION OF WEATHERED ROCK

Grade	Description	Lithology	Excavation	Foundations
VI	Soil	Some organic content, no original structure	May need to save and re-use	Unsuitable
V	Completely weathered	Decomposed soil, some remnant structure	Scrape	Assess by soil testing
IV	Highly weathered	Partly changed to soil, Soil > Rock	Scrape NB corestones	Variable and unreliable
III	Moderately weathered	Partly changed to soil, Rock > Soil	Rip	Good for most small structures
II	Slightly weathered	Increased fractures, and mineral staining	Blast	Good for anything except large dams
I	Fresh rock	Clean rock	Blast	Sound

(More complex schemes, for description of non-uniform and mixed rock masses, are given in BS 5930)

WEATHERING GRADE AND ROCK PROPERTIES

Some representative values for selected materials to demonstrate physical changes in weathered rock

Grade of weathering		I	II	III	IV	V
Granite: unconfined compressive strength	MPa	250	150	5–100	2–15	
Triassic sandstone: unconfined compressive strength	MPa	30	15	5	2	<1
Carboniferous sandstone: rock quality designation	%	80	70	50	20	0
Chalk: standard penetration test	N value	> 35	30	22	17	<15
Chalk: safe bearing pressure	kPa	1000	750	400	200	75
Triassic mudstone: safe bearing pressure	kPa	400	250	150	50	
Triassic mudstone: clay particle fraction	%	10–35		10–35	30–50	
Typical depth in Britain	metres	5–30	1–5	1–2		

WEATHERING PROFILES IN ROCK

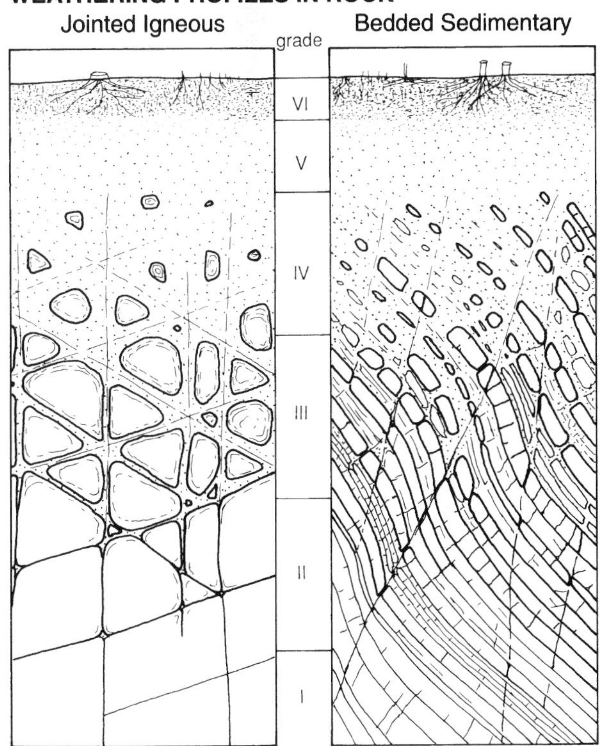

WEATHERING OF LIMESTONE

Limestone is unique because it is a physically strong rock which can be totally removed by solution during weathering.

Rainwater and soil water weather the limestone surface, and also dissolve away the rock where they seep down fractures and bedding planes thereby creating wide fissures and caves.

This process forms very uneven ground with strong rock and large voids.

Pinnacled rockhead has deep fissures, mostly filled with soil, between weathered limestone pinnacles, all beneath soil or drift cover; it creates difficult foundation conditions prone to sinkhole subsidence (section 29).

Limestone pavements with large flat rock surfaces are the result of recent glacial scouring which removed the weathered and dissected surface rock.

Karst is a limestone landscape characterized by underground drainage, caves, sinkholes, dry valleys, thin soils and bare rock outcrops.

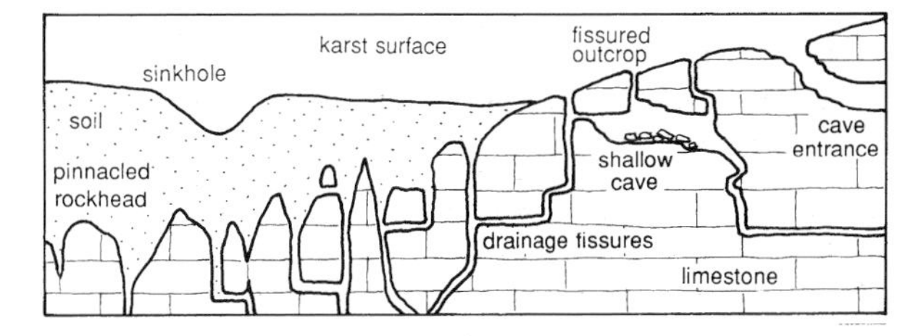

DRIFT DEPOSITS AND CLIMATE

The nature, extent, structure and properties of drift deposits are closely related to the processes by which they were deposited.

These deposition processes are determined largely by climate.

Fluvial processes – the action of rivers and flowing water – are dominant in all climatic regimes except for the permanently frozen zone beneath glaciers and the arid zones in deserts.

Ice Ages: During the Quaternary, the Pleistocene period was marked by phases of worldwide cooling – the Ice Ages – when ice sheets covered large parts of the northern continents, and climates were severely modified across the rest of the world. The last ice sheets retreated only about 10 000 years ago.

Many drift deposits were formed in environments very different from those of today. They are therefore best understood by distinguishing them on the basis of process and climate.

14 Floodplains and Alluvium

WATER EROSION

Water is the main agent of erosion; its power increases greatly with velocity.
Rivers erode by downcutting, and sides degrade to form V-profile valleys. On low gradients downcutting reduces, so lateral erosion dominates, notably on the outside of river bends.
Sediment is transported as rolled bedload and in suspension; particle size increases with velocity. Deposition is due to velocity loss, on gradient loss and inside bends, so sediment is sorted by size.

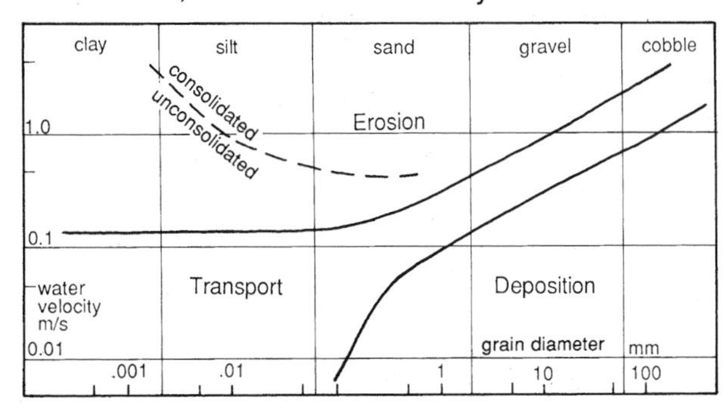

ALLUVIAL DEPOSITS

Alluvium: river-deposited sediment; sorted and bedded, but any grain size from clay to boulders; laterally and vertically variable, with wide range of engineering properties.
Floodplain: zone of alluvial deposition along valley floor, subject to periodic flooding. The alluvium builds up over time, much of it formed as overbank flood deposits which are mostly fine grained and horizontally bedded.
Meander scrolls: cross-bedded, crescentic lenses of sediment, mostly sand or gravel, left on insides of migrating river bends or meanders.
Channel fills: abandoned river channels filled with sediment, commonly clay or peat.
Alluvial fans: coarser, poorly sorted sediment (inc. fanglomerate) on steeper slopes and at mouths of hillside gullies and tributary streams.
River terraces: remnants of any older, higher floodplains, abandoned when river cut to lower level; formed of alluvium, though may be rock-cored; eroded away as modern floodplain enlarges.
Tufa and travertine: weak, porous calcite deposit, forming thin layer or cementing gravel; may overlie uncemented alluvium and can be confused with rockhead. Les Cheurfas dam, Algeria, was built 1885 on tufa crust, and failed by piping on first impoundment.
Peat: black organic soil, formed in small lenses or large areas of upland bog or lowland fen; extremely weak and compressible (section 27).
Lake deposits: similar to fine alluvium (section 15).

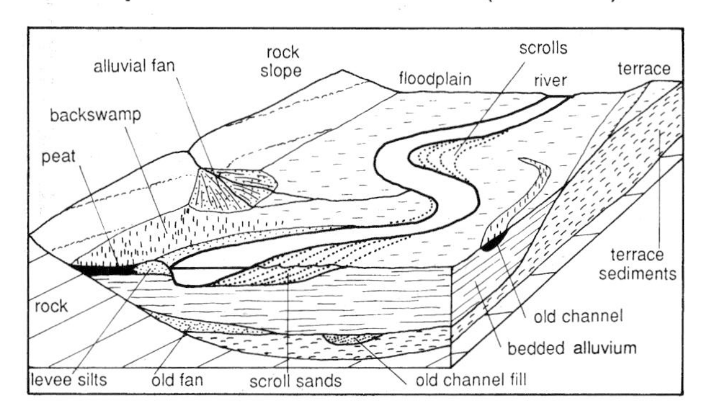

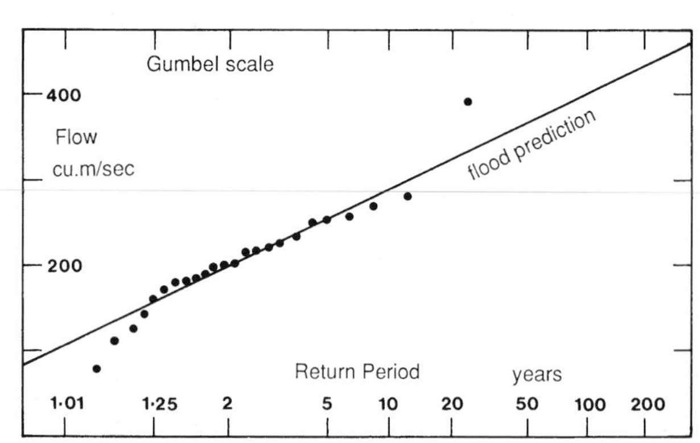

Maximum annual flood flows on a river over 24 years; return period = (number of records – 1)/(rank)

FLOODPLAINS

Flooding is natural and inevitable on floodplains.
Flood size (expressed as flow, stage, height or area) is described by its statistical return period, e.g. a 50 year flood which has 2% chance of occurring in any year.
From existing data, plot of flow against return time (based on rank) gives straight line (often except for the highest flood) which allows predictions of rarer events. So flood zones can be identified and avoided, and channel sizes can be designed.
Floodplain hydrology may be changed unintentionally; urbanization, deforestation and levee construction all raise height of flood peaks.

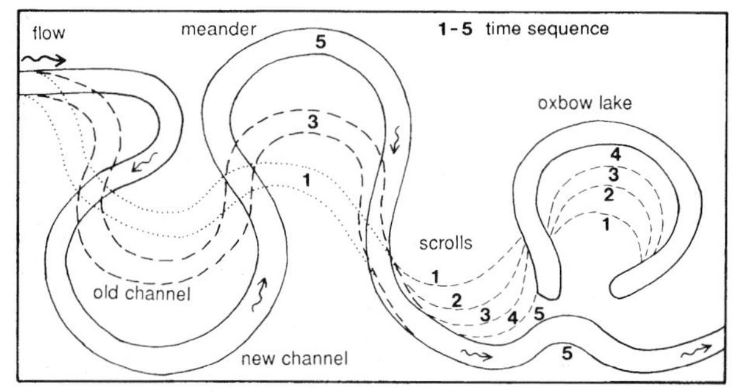

Development of river meanders

CONSTRUCTION ON ALLUVIUM

Alluvium thickness may vary 1–>100 m; difficult to predict but generally compatible with local hill relief.
Some alluvium is laterally uniform. Other has channel fills, scrolls and fans, making site investigation and borehole correlation difficult.
Non-cohesive sand alluvium has SBP = 100–600 kPa, depending on density; quick or running sands form with high water pressures or seepage flows in loose material.
Cohesive clay alluvium generally has SBP = 0–200 kPa, depending largely on consolidation history (section 26).
Bearing capacity of unconsolidated alluvium can normally be increased by effective drainage and consequent accelerated consolidation.

Heavy structures on soft alluvium may require end-bearing piles to rockhead, or friction piles in thicker sequences. Each phase of Yorkshire's Drax power station required over 12 000 end-bearing, pre-cast concrete piles, each 22 m long, driven through clay and silt alluvium to sandstone bedrock or a dense alluvial sand just above rockhead.

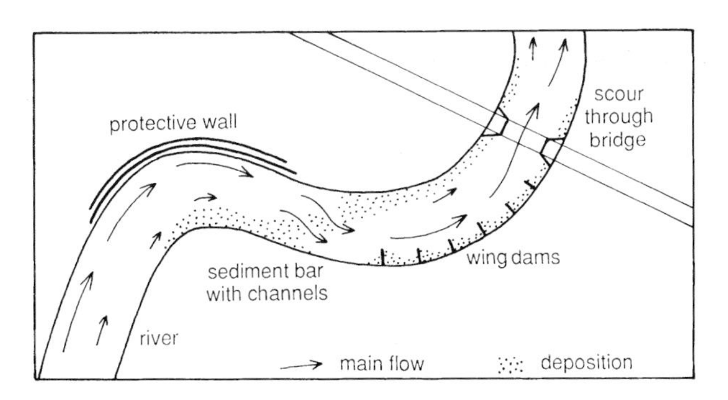

FLOOD CONTROL MEASURES

Levees are linear ridges alongside river channels.
Natural levees are formed by bank overflow and rapid deposition, and may build up to channel a river at a level above the main floodplain; China's Yellow River is 6 m above its floodplain for over 500 km.
Artificial levees are earth (or concrete faced) embankments built to prevent floodplain inundation. They must be continuous – roads must go over them or through floodgates; the Mississippi levees are 10 m high and >1000 km long.
Canalization can shorten a river course, creating a new steeper gradient to transmit flood peaks more effectively.
Levees and canals prevent a river flooding its natural floodplain, and so artificially increase flood peaks downstream. Flood control dams can capture floodwater and act as a substitute for lost floodplain storage.
Floodways are zones of undeveloped land between levees designed to transmit floodwater when required.

CONSTRUCTION ON FLOODPLAINS

Protection by levees permits wider use of floodplains.
On undefended floodplains, construction is normally avoided inside the 100 year flood zone (in Britain the 1947 flood limit provides a useful guideline).
In 1972, the Rapid City flood, in South Dakota, cost 237 lives. All the destroyed buildings were on the floodplain, recognizable on existing maps as the area of alluvium; reconstruction has left the floodplain as a park.
Encroachment is construction on the floodplain which hinders flood flows; it causes upstream ponding, and increased flows and scour round the structures, which are therefore self-destructive and must be avoided.
Parkland and buildings with unenclosed ground floor carparks do not encroach, and are acceptable in active floodplains and floodways.
Transport routes need to cross floodplains; bridges must have extra flood arches to avoid encroachment; flat bridges with no parapets can survive overflooding with no damage and only short-term loss of use.

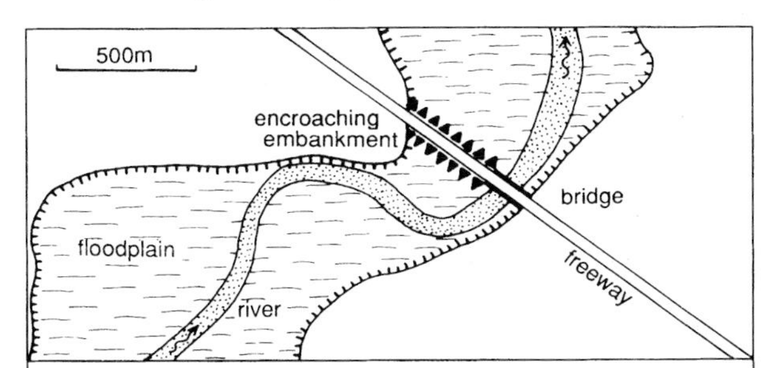

Freeway bridge over Schoharie Creek, New York, failed in 1987; approach embankment had encroached on floodplain and increased flood flow beneath bridge so extra scour undermined pad foundations on gravel.

RIVER CHANNEL ENGINEERING

The natural processes of river flow include:

- Erosion on the outside of bends;
- Channel migration as a consequence of bend erosion;
- Bed scour between encroaching bridge piers;
- Sediment deposition in slack water, notably inside bends and downstream of obstacles;
- Catastrophic channel rerouting across floodplains, during rare flood events which overtop levees.

Bank erosion may exceed 1 m/y and protection may be essential, using walls of concrete or gabions (loose rock in wire baskets), and wing dams to trap sediment.
Repeated dredging may be needed to counter mid-channel deposition, notably on oblique bars.

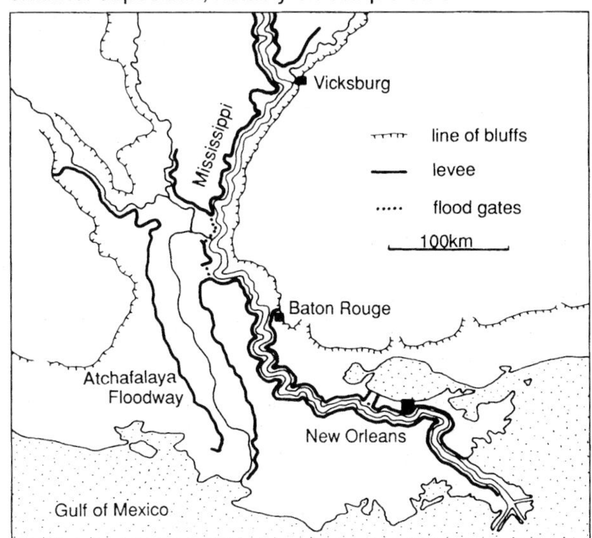

Mississippi River has continuous levees to protect New Orleans and other cities. Floodways sluices can be opened to take flood peaks that threaten to overtop the levees. The 1993 flood overtopped the 100 year levees. Since then, more farmland has been left unprotected – a limit to sustainable economic floodplain development is now recognized.

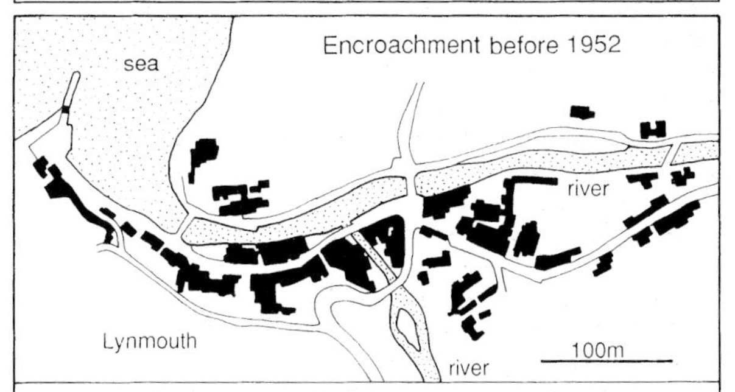

The 1952 Lynmouth flood disaster, in Devon, was due to encroachment by bridges and buildings which diverted floodwater down the streets and through the village. New larger channel has longer bridges and a floodway park.

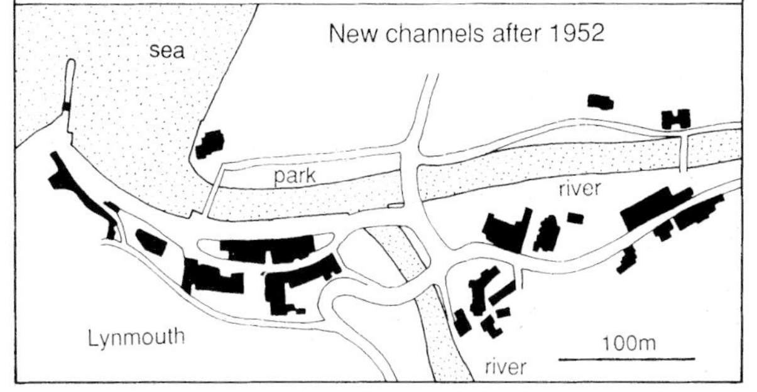

15 Glacial Deposits

Pleistocene Ice Ages

Ice Ages were created by a series of phases of worldwide cooling, when average temperatures fell 8°C.
In each Ice Age, the largest ice sheets centered over Canada (reaching into USA) and Scandinavia (extending over most of Britain). Smaller ice caps formed on most high mountain ranges. Glaciated landforms and deposits still remain over these very large areas.
Last Ice Age ended about 10 000 years ago; known as Devensian in Britain and Wisconsin in USA; features dominate mountain landscapes today, also extensive deposits left on glaciated lowlands.
Earlier Ice Ages had ice sheets more extensive than during the Last Ice Age, leaving lowland deposits in parts of Britain and USA.

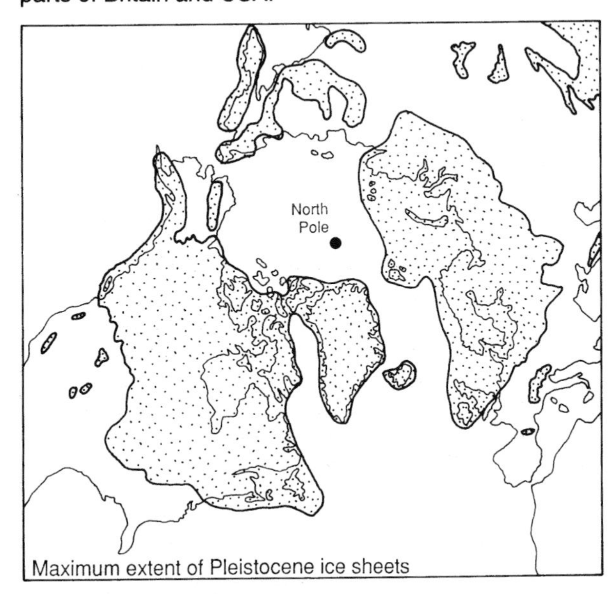

Maximum extent of Pleistocene ice sheets

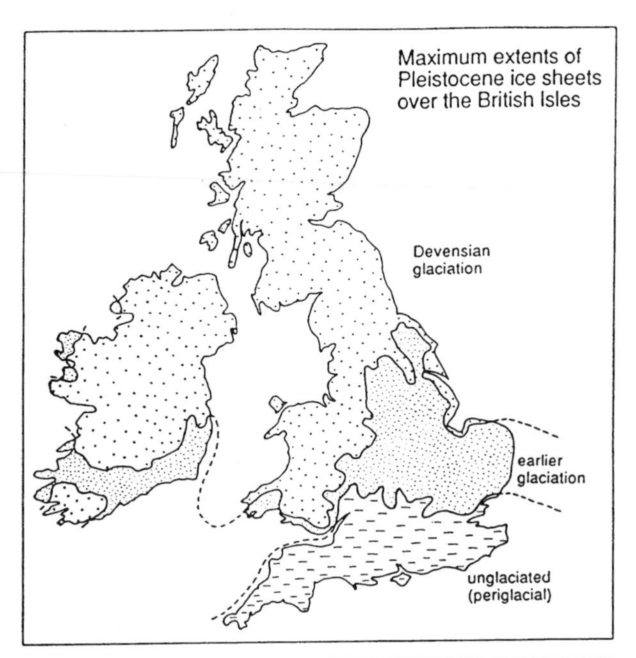

Ice Ages had other far-reaching effects:
Periglacial conditions extended over large areas, including all of southern Britain and the USA as far south as Oregon, Wyoming and Tennessee.
World sea levels fell 150 m as water was locked in ice sheets; Britain was joined to France, Alaska to Siberia.
Weight of ice caused crustal sag beneath ice-sheets, followed by slow isostatic uplift after ice melted.

GLACIAL EROSION

Glaciers form where winter snowfall exceeds summer melt; snow layers accumulate, compressed lower zones recrystallize and are squeezed out as flowing ice.
Most glaciers move about a metre per day.
Pleistocene glaciated areas are most easily recognized by erosional landforms.
Alpine glaciation: valley glaciers in mountain areas of high relief; ice further deepens U-shape valleys between high uneroded arête ridges.
Sheet glaciation: thick ice moves over entire landscape; greater erosion of high ground reduces relief; ice can erode while moving uphill, to create irregular topography with over-deepened rock basins.

GLACIOFLUVIAL DEPOSITS

Meltwater occurs on, in, beneath and downstream of all glaciers; it erodes transports and deposits various types of glaciofluvial sediments.
Sediment is mostly sand and gravel, with moderate sorting and bedding; fines have generally been washed out; commonly non-cohesive and highly permeable, with good bearing capacity and low settlement.
Outwash: tracts of alluvial sand and gravel deposited by meltwater downstream of glacier snouts.
Kames and eskers: hills and ridges which were sediment fills in glacier caves; may be buried inside till.
Glaciofluvial sediments are also known as stratified till, or glacial sand and gravel.

GLACIAL DEPOSITION

Debris of all sizes is picked up and transported by glaciers, and then dumped at glacier edges, along their bases, or mostly in terminal melt zones.
Till or **boulder clay**: general terms for glacial debris.
Moraines: morphological units of glacial till – layers, mounds, ridges or any shape of deposit on, or left behind by, a glacier.
Sheet moraines: extensive till blankets of variable thickness; typically hummocky surfaces may be streamlined into drumlin landscapes.
Valley moraines: till ridges along or across glaciers or left behind in glaciated valleys:

- lateral moraines: along glacier edges, fed by debris from higher slopes;
- medial moraines: coalesced lateral moraines where glaciers converge (not so common);
- terminal, end or retreat moraines: till banks across valleys where ice melting reaches maximum at the glacier snouts; a sequence may be left up a valley by a glacier retreating irregularly.

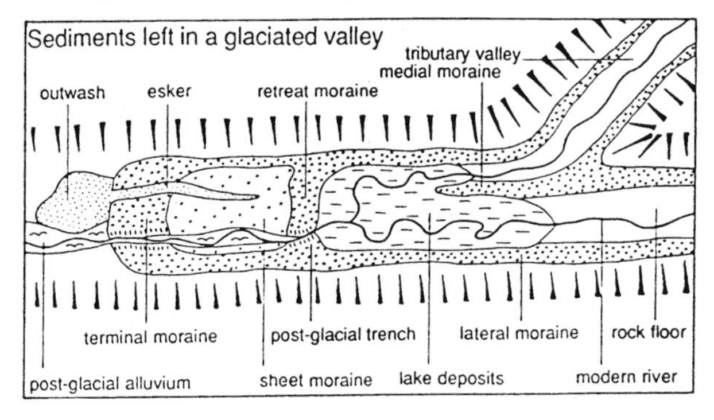

BURIED TOPOGRAPHY

Rockhead relief: features of an eroded landscape buried beneath drift.

Burial is consequence of deposition after erosion – common with Pleistocene climatic changes.

Topography of old buried landscape (rockhead) may not relate to modern landscape.

Depth to rockhead may be variable and irregular; greatest and least predictable under post-glacial drift in over-deepened glaciated valleys and where subglacial meltwater channels entrenched bedrock beneath ice.

Buried valleys have drift of locally greater or unknown thickness, causing added costs and potential hazard for foundations designed to bear on rockhead. May lie beneath, or be unrelated to, modern valleys. Irregular subglacial channels in rockhead commonly have very steep sides; often known as tunnel valleys.

GLACIAL TILL

Till is unsorted and unstratified glacial sediment consisting of a mixture of any or all of clay, silt, sand, gravel, cobbles and boulders.

Composition relates to the rocks which were eroded by glacier before deposition.

Also known as boulder clay – but this term can be misleading as a till with a sandy matrix may have no clay component.

- Lodgement till: carried and deposited at base of ice; generally over-consolidated by overriding glacier, and with clay content of 10–40%
- Ablation till: deposited as ice melted from beneath it; poorly consolidated, commonly with clay content of <10%, as fines removed by meltwater.

All till may be locally variable, with lenses or zones of soft clay, running sand or large boulders. Terminal moraines may be structurally complex where glacial readvance has pushed till into ridges.

Bearing capacity may vary from 400 kPa for old, stiff lodgement till to <100 kPa for ablation till.

Compressibility is generally low except for clay-rich ablation till.

- Excavation costs on St Lawrence Seaway, on USA–Canada border, doubled when dense lodgement till was found instead of loose ablation till (which was at outcrop and wrongly expected at depth).

Temporary cut faces may be vertical in cohesive lodgement till, but need support in sandy ablation till.

Permeability is generally low but variable, related to matrix. Eigiau Dam in North Wales failed in 1925 due to piping through a sandy zone in foundation till.

Erratics: isolated large boulders; may exceed 10 m diameter; may be confused with rockhead in site investigation. Test bores for Silent Valley Reservoir, Ireland, stopped at rock at –15 m, but all had hit erratics; rockhead was –60 m. Driven piles and sheet piling cannot be used in till with erratics.

Glacial till in a Derbyshire quarry

LÖTSCHBERG TUNNEL DISASTER, 1908

Swiss tunnel heading drove through rockhead into saturated gravels 185 m below valley floor, after false assumption of sediment depth. No allowance for any reverse gradient on rock floor of glaciated valley buried beneath alluvial fill. In reality, prediction of rock profile was impossible without boreholes. Horizontal probes ahead of tunnel face would have provided warning. Inrush killed 25 men; tunnel was rerouted.

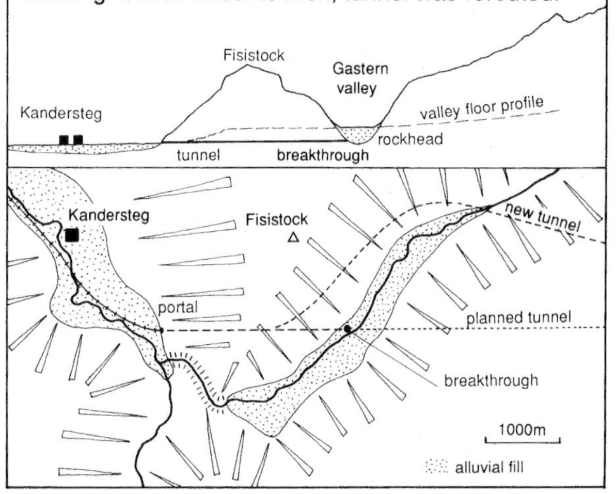

KNOCKSHINNOCH MINE DISASTER, 1950

Scottish mine heading broke through rockhead into glacial till beneath hollow filled with saturated peat on hummocky sheet moraine. Piping failure of till allowed peat inrush to mine, leaving surface sinkhole 100 m across. Flat area on ground profile indicated some sort of fill – and potential hazard; needed checking before heading was advanced to rockhead.

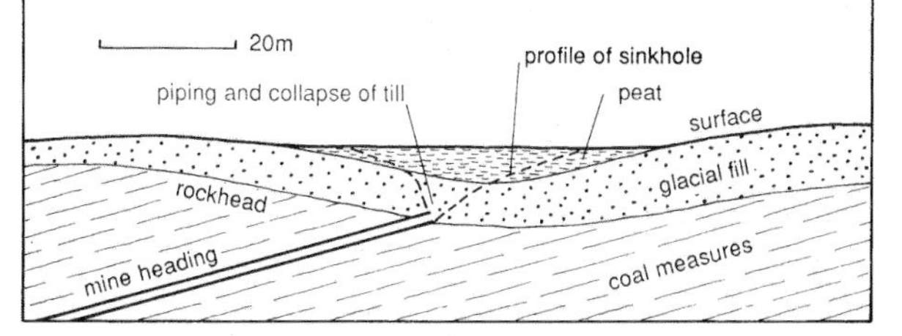

LAKE SEDIMENTS

Most lakes are created by glacial processes – damming behind terminal moraines, and post-glacial flooding of ice-scoured rock basins, over-deepened glaciated valleys and hollows on sheet moraines.

Lakes of English Lake District are in glaciated valleys with end moraines. Great Lakes of America are in ice scoured basins, partly dammed by moraines and ponded by post-glacial isostatic uplift of outlets.

Rivers destroy lakes – by sediment infill at upper end draining and by erosional lowering of outlet. Thousands of lakes left at end of Pleistocene have since been filled and/or drained, leaving areas of lake sediments.

Lake (lacustrine) sediments are like alluvium, generally with more silts and clays, and less gravels so commonly have lower bearing capacity with higher settlements.

Often recognize lake sediments by flat ground.

Small ponds in sheet moraine commonly filled with mosses to form peat.

Sensitive clays were deposited in inland seas along Pleistocene ice margins in Scandinavia and eastern Canada (section 34).

16 Climatic Variants

Semi-Arid Environments

Deserts have low rainfalls – less than potential evaporation – and may be hot or cold. With the lack of surface water, wind erosion and transport become effective but periodic water erosion is still the dominant process, except in rare totally arid deserts.

Wadis, or arroyos, are desert valleys, normally dry but subject to flash floods from isolated rainstorms. Flood flows decrease and sediment loads increase downstream. Roads across wadi floors which only rarely flood can be designed to be overflooded; built on gabions to stop downstream scour and undercutting.

Selective erosion is by slow weathering, wind transport and isolated flood events; leaves residual inselberg mountains, flat-topped mesas and pillar buttes in layered rocks, and natural arches where weathering breaches thin rock ribs.

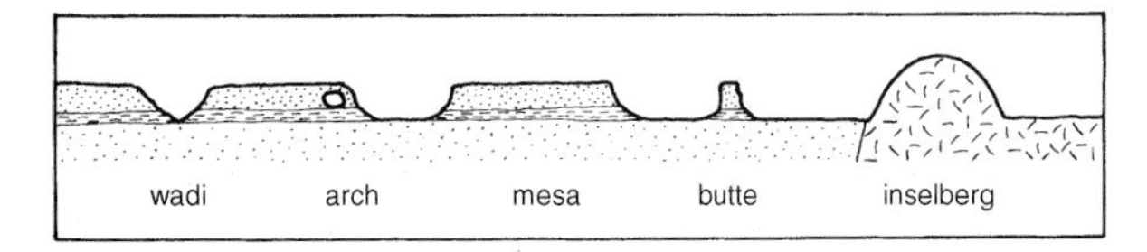

Desertification is the slow loss of vegetation and expansion of the desert, as in the African Sahel. It is due to any or all of climatic change, deforestation, overcropping, overgrazing, and soil salt increase by evaporation of irrigation water.

BLOWN SAND

Wind moves dry sand by sliding, rolling or bouncing (saltation). Sand abrasion undercuts rocks and structures close to ground level. Deflation removes sand leaving desert pavements of polished pebbles.

Dunes are built by deposition in slack air, in eddies and in the lee of obstacles; they may be longitudinal (seif) or transverse to prevailing wind, irregular in shape or crescentic barchans. Active, depositional slopes of dunes have loose sand at angle of repose of 32–34°, while the flatter sides are eroded by the wind in firm dense sand.

Stabilization of migrating dunes may be achieved by trapping the sand with induced vegetation cover, or fences and shelter belts; but impractical if sand supply is too large. Sand accumulates downwind of structures, while exposed areas are cleared by the wind; self-clearing roads must slope gently upwind.

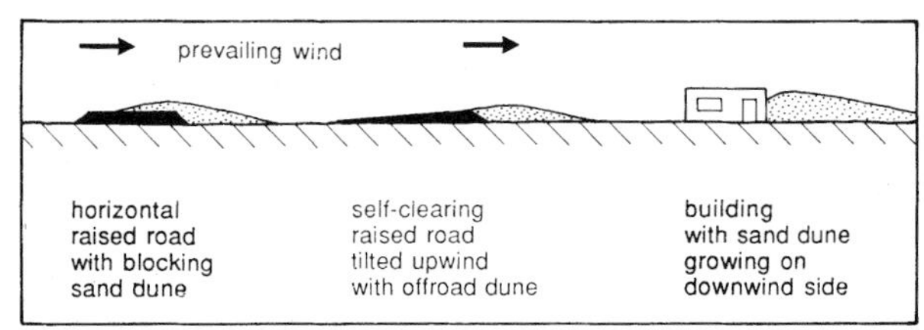

Corrugations on dirt roads develop on gap-graded material (*eg* silty sand) as the fines squeeze into voids between the larger grains. Well graded soil is stable.

ARID LANDFORM ZONES

	zone	slope	width	process	sediments	drainage	hazards
1	Mountain	>12°		erosion	(rock)	gorges	flash floods
2	Pediment	2–12°	1–2 km	steep fans	coarse unsorted	entrenched wadis	hydrocompaction
3	Alluvial Plain	0·5–2°	1–>10 km	gentle fans	sand, gravel, fines	shallow wadis	and blown sand
4	Playa	< 0·5°		basin flat	silt, mud, salt	temporary lakes	salt, blown sand

DESERT SEDIMENTS

Alluvial fans: banks of flood sediment and debris flows from mountain wadis. May coalesce into bajada, forming the mountain footslope with a sediment apron over sloping bedrock pediment. Sediment is rapidly deposited, unsorted and poorly consolidated.

Alluvial plains: extensive lower reaches of fans; mainly sand and gravel in shallow, braided wadi channels; coarse cobble beds commonly remain from wetter Pleistocene climates.

Playas: flat floors of inland basins with temporary evaporating lakes. Mostly fine, soft, weak, silts and clays, often thixotropic, with salt and gypsum evaporites. Coastal sabkha zones are similar.

Salt: may form thick beds, with other evaporites in playa and sabkha zones; also left by evaporation as component of clastic sediments. Capillary rise in fine soils may lift salt 3 m above water table into road and building structures. Salt crystal growth is major form of desert weathering – of rocks and concrete. Dense, impermeable concrete suffers less from salt breakdown.

Duricrusts: surface layers of cemented sediment, mostly sand or gravel; mineral cement deposited by evaporating groundwater. Most common duricrust is calcrete, or caliche, cemented by calcite, about 1 m thick, over unconsolidated sediment; should not be confused with rockhead as bearing capacity is low.

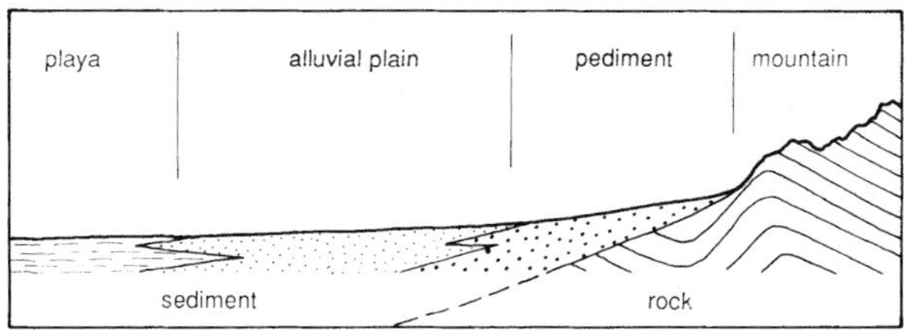

Profile through the typical desert landform zones

COLLAPSING SOILS

These are mainly the fine grained, low density soils that restructure and compact due to the addition of water, a process known as hydrocompaction.

Loess is structureless, yellowish, calcareous silt, of grain size 0·02–0·06 mm, common in the interiors of the northern continents; much of it was derived by wind deflation from Pleistocene glacial outwash plains. Dry or moist loess will stand in a vertical face; but it is easily gullied and piped by running water, and disaggregates and collapses on saturation.

Alluvial silts, if rapidly deposited by floodwaters, and then dessicated, on the lower parts of fans, may also hydrocompact; common in the semi-arid basins of inland California.

Periglacial Environments

Cold climates with limited snowfall are periglacial, with no ice cover; locally peripheral to glaciers. Winter snowfall is matched by summer melt, and vegetation is sparse tundra.

Today, much of Alaska, Arctic Canada and Siberia is periglacial; in Pleistocene, conditions extended over southern Britain, central Europe and central USA, south of the ice sheets, leaving periglacial structures which are found in ground today.

Climate causes increased frost shatter and mechanical weathering, and reduced chemical weathering.

Permafrost: permanently frozen ground, to depths of 10– > 200 m; deeper freezing is prevented by geo-thermal heat; if permafrost < 50 m thick, it may be discontinuous; may include lenses of pure ice within soil and rock.

Active layer: zone of summer melting and winter refreezing, generally 0·3–6 m deep; frozen ground beneath prevents drainage, leaving it saturated and unstable in summer, causing widespread slope failure and subsidence.

STRUCTURES AND SEDIMENTS

Landslides and solifluction common in active layer.

Camber folding, valley bulging increase (section 06).

Ice heave and collapse forms irregular cryoturbated ground, sediment-filled ice wedges, patterned ground with stone polygons; all create disturbed and vertical boundaries in soil active layer. Deeper drift-filled hollows in London Clay relate to freezing around artesian groundwater flows.

Frost shatter is extensive, commonly to 10 m deep in chalk of southern England.

Scree, or talus, is coarse, angular slope debris, with angle of repose equal to 37°, masking cliff foot profile. Many fossil Pleistocene screes are inactive in modern climate and so gain vegetation cover.

Lowland periglacial sediments include outwash gravels, alluvial and blown sand, and extensive loess.

Clay with flints is soliflucted mixture of residual soils and Tertiary clastics, widespread but thin on English chalk outcrops.

SOLIFLUCTION

This is the downslope movement of saturated debris – a type of wet soil creep moving about 1 metre per year. It can occur on any saturated slope, but is most common in the summer-thawed active layer of periglacial slopes which cannot drain through the deeper permafrost.

Head is unsorted, soliflucted debris; it may appear similar to glacial till, but is formed entirely of local upslope material. Head can flow by plastic deformation, but is typically well sheared, with basal, intermediate and circular slip surfaces. The shears reduce the strength to low residual values.

Solifluction flows, up to 1000 m long, may move on slopes as low as 2°; commonly 2–4 m thick, but may accumulate in layers to depths of >15 m on concave slopes and as valley infills.

Head forms most easily on slopes of shale, mudstone, clay and chalk; coombe rock is chalk head. Most slopes in Britain, steeper than 5° on these rocks and outside the Devensian ice limits, can be expected to have a veneer of sheared, unstable head.

Carsington Dam, an earth embankment in Derbyshire, failed in 1984 before the reservoir was filled. A slip surface developed through both the weak clay core (of unusual shape) and a layer of head left on the shale bedrock beneath the placed fill. The head was wrongly interpreted as in situ weathered shale, and the design assumed an undisturbed angle of friction $\phi = 20°$, but shear surfaces reduced its strength to a residual $\phi = 12°$. This mistake, and the rebuild, cost £20M, yet periglacial head is widespread on the shale outcrops of Derbyshire and could have been expected.

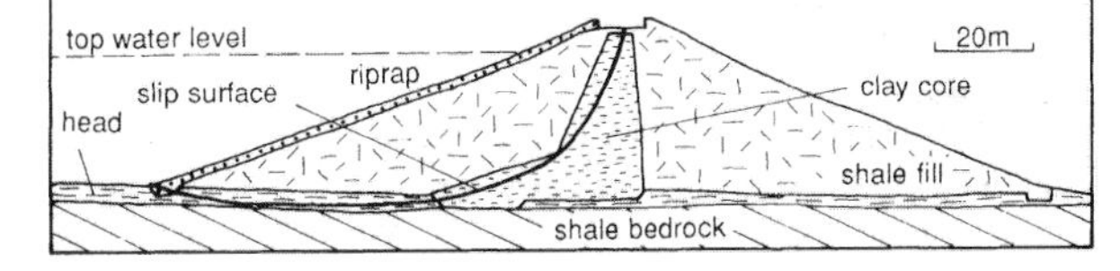

PERMAFROST ENGINEERING

Subsidence, flow and heave occur on poorly drained silts and clays when ground ice is melted; sands and gravels are generally thaw-stable.

Conservation of the permafrost is generally best. Any disturbance of natural insulation (by soil and vegetation) increases summer thaw and depresses permafrost beneath buildings and roads.

Block supports for heated buildings, with clear airspace beneath, can be stable on gravel active layer over preserved permafrost.

Piles into stable frozen ground generally need to reach depths around 10 m.

Utilidors are pile-supported conduits built in streets for heated services.

Alaska pipeline rests on piled trestles, each with internal circulating coolant and heat fins on top to dissipate stray heat from the oil.

Gravel pads or embankments, a few metres thick, can be enough to provide insulation and let the permafrost expand into them, stabilizing the compacted old active layer. Internal cold air ducting or insulation layers of peat or wood chips further improve permafrost protection.

Subsided houses on permafrost in Dawson, Canada

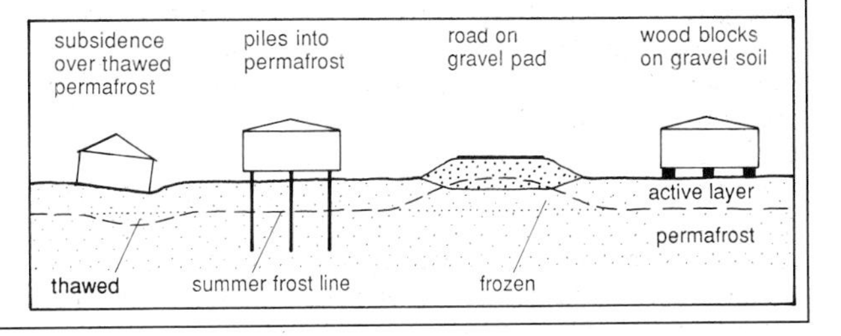

17 Coastal Processes

Wave action is the dominant mechanism in both erosion and deposition along coastlines.
Waves are powerful due to their pounding action but are also very selective in etching out rock weaknesses and controlling sediment deposition.
The largest and most powerful waves are those which have travelled furthest, i.e. have the greatest fetch.
Large storm waves are powerful and destructive.
Tidal surges cause damage when waves reach new heights; they occur on high spring tides, aided by strong onshore winds and low atmospheric pressure, as in the North Sea, 1953, and at Bangladesh, 1970.

CLIFF EROSION

Coastal erosion is by wave action at beach level.
This creates a wave notch, which advances, leaving a wave cut platform, and undercutting the cliffs – which retreat by a sequence of rock falls (or larger landslides in weak material).
Selective wave erosion of strong rock cliffs etches out faults, joints and weaknesses to form sea caves, arches and inlets which retreat between headlands and sea stacks.
Coastal equilibrium produces slowly eroding headlands between bays where soft rocks are protected by beach deposition.

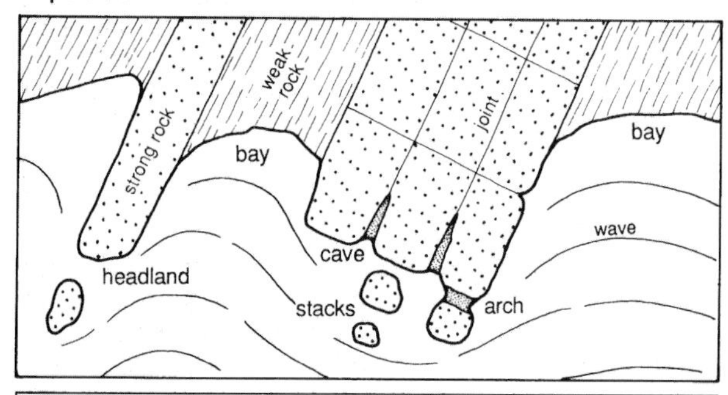

SHORELINE BUDGET

The main source areas of beach sediment are eroding cliffs of soft Pleistocene material, and also active river delta deposition.
On any shoreline, erosion and the production, transport and deposition of sediments are all finely balanced; any disturbance of the budget prompts renewed erosion or deposition to recover the equilibrium.
The longshore drift budget is easily disturbed by coastal engineering works, most notably where sediment is trapped or deflected into deeper water, so that downdrift beach starvation causes renewed erosion.
Similarly, any disturbance of a beach profile prompts natural processes which oppose the change in order to restore a stable form.

Coast erosion at Holderness, eastern England

SEDIMENT TRANSPORT

Clastic sediment is mostly rolled along the seabed where it is reached by wave motion in shallow water.
Beaches are formed by sand deposition where wave upwash (swash) is greater than the backwash due to water soaking down into the porous sand.
Shingle storm beach forms higher up by larger waves.
Coastal dunes are of beach sand blown inland by wind.
Sediment washed into deeper water is deposited below wave influence, as wave-built terrace or offshore bar.
Longshore drift is due to oblique upwash of impacting waves, then backwash directly down beach slope – always away from waves arriving with greatest fetch.

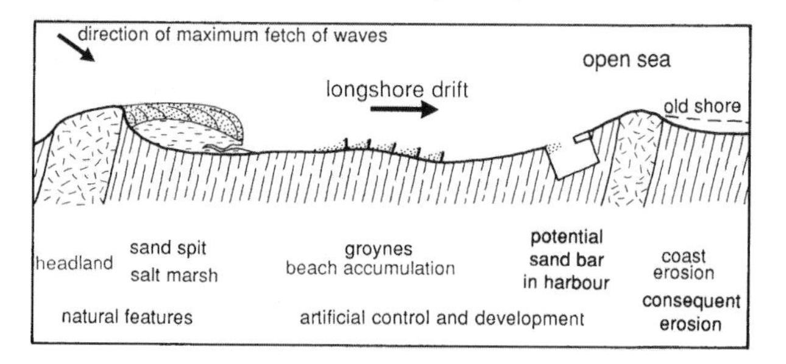

Deposition occurs in the lee of any obstacle, where sand is drifted into deep or slack water; a spit forms downdrift of a headland, and may extend into a bar across a bay or river mouth; a tombolo may form a sand link between an island and the main shore.
Salt marsh forms on mud held behind spits, storm beaches or coastal dunes.
Barrier islands, as on east coast of USA, may evolve from extended spits or emergent offshore bars.
Carbonate sediments in shallow water create varied and difficult ground conditions for coastal and off-shore structures. Strengths of shell sands and coral reefs vary greatly with type and cementation history.
Lagoons and sabkha sediments include muds, gypsum and carbonate weakened by karstic solution cavities.
Western Australia off-shore oil platforms required piles through more than 110 m of loose shell sands.

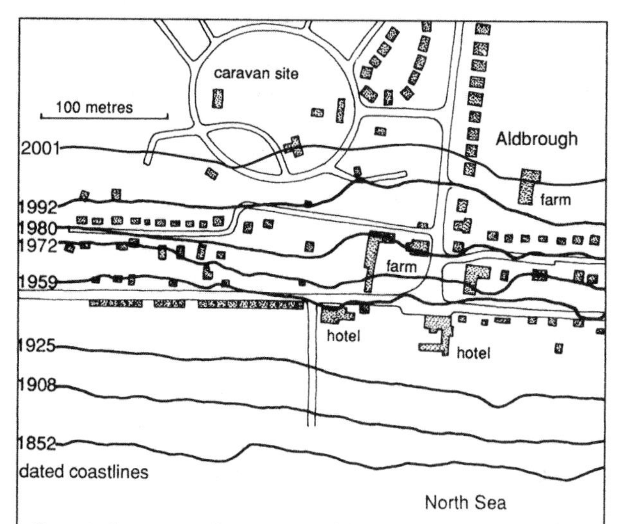

Coastal retreat is most rapid where soft rock cliffs are eroded and longshore drift leaves no protective beach. On the east coast of England, the glacial till cliffs of Holderness retreat about 2 m/year, and successive landslides destroy houses, farms and roads.

Coastal Engineering

EROSION DEFENCES

A wide sand beach is the best means of preventing coastal erosion and retreat. Sea walls may prevent erosion, but wave action is relentless, and even the largest structure is only a short-term defence unless there is effective beach sediment control.

Efficient sea defence is porous to absorb wave energy; made of armour stone (blocks of >2 tons), concrete tetrapods, or massive wall faced with cellular concrete. Reflected waves off solid face may induce scour.

Sea walls may cost £5M/km. Economical alternative on long eroding coast (eg Holderness) is to create hard points – short sections of stable, fully defended shore – with intervening coast left unprotected. Down-drift of each hard point, erosion creates a shallow bay, which traps beach sediment. Eventually, a crenulated coast should become stable, but compensation is needed for short-term accelerated land loss between hard points.

BEACH CONTROL

Groynes are timber, concrete or steel barriers across beach which prevent or reduce longshore drift by trapping sediment. Groyne spacing should be double their length to effectively stabilize beach.

Offshore breakwater, parallel to shore, absorbs wave energy and causes beach accumulation in its lee – similar to on a natural tombolo.

Beach may be stabilized or expanded by pumping seawater from a buried porous pipeline. Wave upwash adds sand to foreshore, but a drained beach absorbs and reduces wave backwash – so that sand is not swept back out to sea.

Active spits, bars and barrier islands migrate inland mainly by wave overwash. Any development, with erosion defences on the exposed outer face, causes thinning due to continued sediment loss from the inner face. The Spurn Head spit, England, and the Carolina barrier islands, USA, are now precariously thin; they should be allowed to break up and reform at a stable site further inland, as artificial defences will become increasingly expensive.

CHANNELS AND HARBOURS

Harbours, cut into the coastline or built out between breakwaters, are stable on a coast which is an erosional source area of overall sediment losses.

Harbour mouths may develop obstructing sand bars if longshore drift is strong. Jetties deflect sediment drift; they may develop spits off their ends and cause downdrift beach starvation.

Natural clearance of harbour and lagoon channels relies on tidal scour, which must exceed deposition by beach drift; larger tidal volumes and flow velocities improve scour clearance, so larger lagoons and narrow channels are better kept clear.

TSUNAMIS

These are large waves generated by seabed earthquake movements; they form in series of 1–6 waves. In the open ocean they are long and low, but they slow down in shallow water, and can build up to >10 m high approaching a shoreline; they reach maximum heights in tapering inlets.

Most tsunamis occur in the Pacific Ocean, and take up to 24 hours to travel from the earthquake location to distant shores. The practical defence for such rare events is warning and coastal evacuation; the Pacific is covered by an efficient international warning system.

BEACH STARVATION

Sediment input and output, by longshore drift, must be in balance to maintain a stable beach. Many artificial measures – trapping drift on a groyned beach, reducing erosion with a sea wall, deflecting sediment at a harbour mouth – reduce onward drift, and therefore cause beach starvation at downdrift sites.

This may cause beach loss or renewed erosion (as at Folkestone Warren, section 36) at new sites downdrift of engineered sections. Beach nourishment by artificial input of sediment is an expensive alternative to downdrift extensions of the initial control measures.

Hallsands village stood on a rock platform with a protective beach in front of it, on the Devon coast. In 1897, offshore shingle dredging steepened the seabed sediment profile. Natural response was lowering and removal of beach within five years; so houses were exposed to waves, and destroyed in a storm in 1917.

SEA LEVEL CHANGES

Pleistocene sea levels fell by about 150 m when water was trapped in continental ice sheets, and some land areas were depressed as much as 50 m by ice weight.

Drowned valleys (rias) were flooded by sea level rise at the Ice Age end, after having been cut by rivers draining to the lower sea levels; some now form natural harbours, as Milford Haven and Plymouth; others have sediment fills, leaving deep coastal buried valleys.

Raised beaches have abandoned cliffs, dry sea caves and fossil beach sediments; many were cut in ice-depressed coastlines at end of Ice Age after sea level had risen but before land had isostaticly rebounded; Scotland's raised beaches are due to its Pleistocene ice burden; California's are due to plate boundary uplift.

Unconsolidated raised beach sediments may be clays, sands and/or gravels, typically with lateral variation.

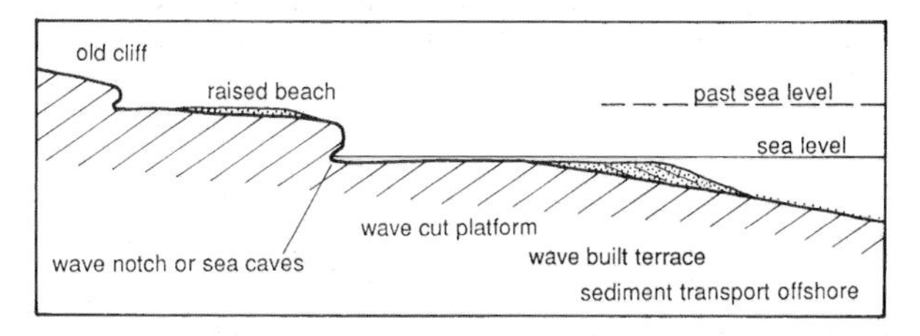

Modern sea level rise is about 1200 mm/100 years worldwide, due to glacier melting which may increase with artificial global warming. Local tectonic movements may greatly increase or reduce the local effect.

Rising sea levels, or ground subsidence, accelerate coastal erosion, cliff retreat, coastal flooding, beach losses and barrier island migration. Greatest effect is on low eastern coastlines of both Britain and USA.

18 Groundwater

Rainfall (precipitation) is the ultimate source of all fresh water, and when it lands on the ground surface it is dispersed in three ways:
Evapotranspiration: combination of evaporation from open water and transpiration by plants, both returning water to the atmosphere; in temperate climates it may vary from 20% of the rainfall on open hills to 70% from wooded lowland.
Runoff: surface water flow into streams and rivers; increases with low rock permeability, steep slopes, intense rainfall and urbanization.
Infiltration: seepage into the ground to become groundwater; important on permeable rocks, and where runoff is slow.
Groundwater is all water flowing through or stored within the ground, in both rocks and soils; it is derived from infiltration, and is lost by flow to surface springs and seepage out through the sea bed.
Water budget is the balance of flows for any part or the whole of a combined groundwater and surface water system; a natural budget is easily disturbed by man's activities, notably where land drainage or urbanization reduce infiltration and groundwater recharge.

PERMEABILITY OF ROCKS

Permeability is the ability of a rock to transmit water through its interconnected voids.
Aquifer: rock with significant permeability, suitable for groundwater abstraction, e.g. sandstone.
Aquiclude: impermeable rock with static water held in poorly connected voids, e.g. clay.
Aquifuge: impermeable rock with no voids, e.g. unfractured granite.
Aquitard: rock with very low permeability, unsuitable for abstraction but significant in regional water budgets, e.g. siltstone.
Permeability (= hydraulic conductivity = coefficient of permeability = K) = flow through unit area of a material in unit time with unit hydraulic head. K is expressed as a velocity, correctly as metres/second, more conveniently as metres/day (in America as Meinzer units = gallons/day/ square foot = 0·0408 m/day).
Intrinsic permeability (k), expressed in darcys, is also a function of viscosity, only significant in considering oil and gas flows through rock.
Groundwater velocities are normally much lower than the K values because natural hydraulic gradients are far less than the 1 in 1 of the coefficient definition. Typical ground-water flow rates vary from 1 m/day to 1 m/year, but are far higher through limestone caves.
Porosity: % volume of voids or pore spaces in a rock.
Specific yield: % volume of water which can drain freely from a rock; it must be less than the porosity, by a factor related to the permeability, and indicates the groundwater resource value of an aquifer.

Typical hydrological values for rock

	Permeability m/day	Porosity %	Sp. Yield %
Granite	0·0001	1	0·5
Shale	0·0001	3	1
Clay	0·0002	50	3
Sandstone (fractured)	5	15	8
Sand	20	30	28
Gravel	300	25	22
Limestone (cavernous)	erratic	5	4
Chalk	20	20	4
Fracture zone	50	10	

K < 0·01 m/day= impermeable rock
K > 1 m/day = exploitable aquifer rock

AQUIFER CONDITIONS

Water table (= groundwater surface) is the level in the rocks below which all voids are water-filled; it generally follows the surface topography, but with less relief, and meets the ground surface at lakes and most rivers.
Vadose water drains under gravity within an aerated aquifer above the water table.
Phreatic water flows laterally under hydrostatic pressure beneath the water table; it is the resource for all high-yield wells; there is less at greater depths and pressures, and most rocks are dry at depths > 3 km.
Capillary water rises above the water table by surface tension, by very little in gravels, by up to 10 m in clays.
Hydraulic gradient is the slope of the water table, created by the pressure gradient necessary to overcome frictional resistance and drive the phreatic flow through the aquifer rock. Water table is steeper where permeability is low or flow is high; typical gradient is 1:100 in good aquifer. Groundwater flow is in direction of water table slope, identified in unpumped wells.
Rivers normally have water table sloping towards them, with groundwater flow into them. Ephemeral rivers lie above water table, and leak into the aquifer.
Perched aquifer lies above the regional water table.
Unconfined aquifer has vadose zone in upper part.
Confined aquifer has artesian water held beneath an overlying aquiclude, with a head of artesian pressure to drive the water above the aquifer, perhaps to rise to ground level; artesian water is common in alluvial sand-clay sequences and in complex landslides.

Groundwater flow = Q = Kbwi, where K = permeability, b = aquifer thickness, w = aquifer width and i = hydraulic gradient. This is Darcy's law, easily calculated for a simple geological structure or as a rough guide for flow through a cut face; the maths is more complex for convergent flow to a well or spring where the water table steepens to compensate for the decreasing cross-sectional area of the aquifer.

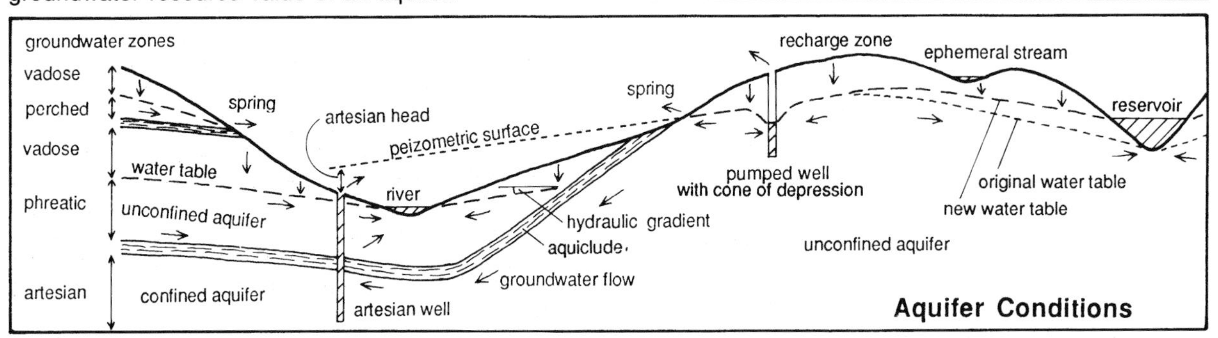

Aquifer Conditions

TYPES OF PERMEABILITY

Intergranular: diffuse flow, between grains, in sands and gravels, poorly cemented sandstones and young porous limestones.
Fracture: through joints, in nearly all rocks; erratic flow in fault zones, but dense joint systems provide diffuse flow in sandstones, chalk and young basalts; most fractures are tight at depths >100m.
Secondary: groundwater flow increases permeability by solution, notably in limestones; non-diffuse conduit flow is erratic through enlarged fissures and caves.

GROUNDWATER DEVELOPMENT

Springs are natural groundwater overflows from aquifers; many are capped or ponded for supply; a large spring yields 0·1–1·0 m^3/s; smaller springs are used in rural areas; limestone caves may feed larger springs.
Qanats are ancient, horizontal adits hand-dug to a sloping water table and freely draining to the surface.
Wells are hand-dug or drilled to below the water table; hand-dug wells may have horizontal adits to intersect productive fracture zones; wells need pumping unless they are artesian; well yield depends on depth below water table, diameter and aquifer permeability; a good well yields 0·1 m^3/s, or about 3 litres/s/m depth below water table; improve yield by blasting to raise fracture permeability near well, or acid injection in limestone.
Cone of depression in water table is formed where pumped flow converging on a well creates steepening hydraulic gradient; the depth of the cone is the well drawdown, related to permeability and flow.
Reservoir impoundment raises the local water table; groundwater leaks through a ridge if water table slope is reversed in an aquifer that reaches a nearby valley.

Pump testing of a well determines its potential yield, and also the regional permeability of the aquifer.

$$K = Q.\ln(B/A)/\pi(b^2 - a^2)$$

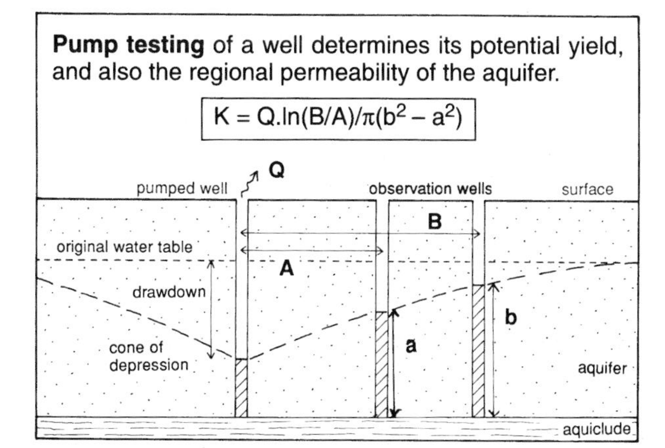

KARST GROUNDWATER

Cavernous limestones do not conform to normal groundwater rules because caves carry water in erratic and unpredictable patterns.
Limestones have complex water tables unrelated to topography.
Karst groundwater is difficult to abstract or control, as wells and boreholes can just miss major conduits.
Cave streams transmit undiluted pollution to springs.

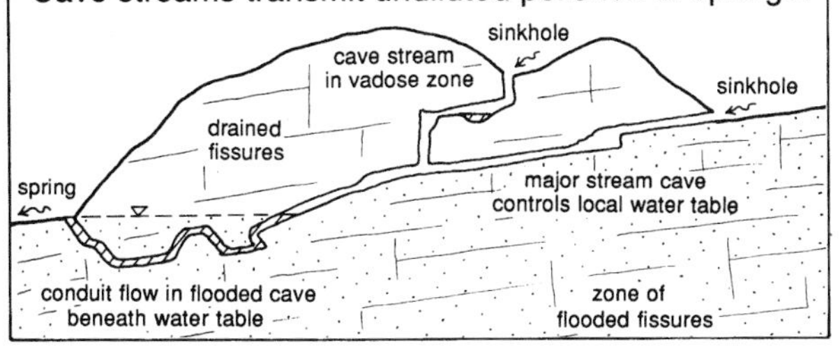

PORE WATER PRESSURE

The groundwater head provides the pore water pressure (p.w.p.) in saturated rocks and soils.
Increased p.w.p. may cause slope failure (section 33)
Decreased p.w.p. may permit or cause subsidence in clays (section 28).
In fractured rocks, joint water pressure is equivalent to p.w.p. and is critical to slope stability (section 32).

GROUNDWATER CONTROL

Dry excavation below the water table is possible within coalesced cones of depression from pumped well points round a site perimeter.
Groundwater barriers permit dry excavation without lowering the surrounding water table; barriers may be steel sheet piles, concrete diaphragm walls, grouted zones or ground freezing, in order of rising cost; grouting or freezing can also control rising groundwater in thick aquifers.
Slopes may be drained by ditches, adits or wells.
Capillary rise in embankments is prevented by a basal gravel layer.

Packer test measures local permeability of rock and aquifer properties between two inflatable packer seals in a borehole.

$$K = Q.\ln(2L/D)/2\pi LH$$

H is measured to water table or to midpoint of test zone if this is above water table. Falling head test is better for low permeabilities.

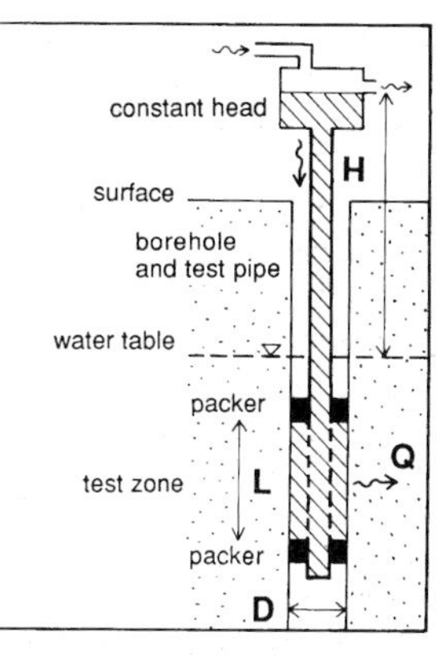

GROUNDWATER RESOURCES

Aquifer stability only ensured if abstraction < recharge.
Abstraction > recharge is groundwater mining – aquifer is depleted; water table falls, springs and wells may dry up, pumping costs increase, artesian wells may cease to flow, resource will ultimately be lost.
Aquifer recharge is possible through intake wells or leaky reservoirs.
Artesian water emerges unpumped from a flowing artesian well. Large resources may lie in synclines.
Groundwater quality is ensured by aquifer filtration and the underground residence time in contact with absorptive clays and cleansing bacteria in soils.
Pollution is most likely in shallow alluvial gravels and cavernous limestones; major pollutants are tank leaks, and hydrocarbons from road drains in recharge zones.
Water hardness is carbonate (limestone) and sulphate.

Villa Farm disposal site, near Coventry, separated liquids in lagoons in old sand quarry 50 m across. Fluid loss of 7000 m^3/y was infiltration to sand aquifer. Pollution had little radial spread, but formed plume 600 m long in direction of hydraulic gradient.

Saltwater intrusion near a coastline is caused by overpumping which disturbs the saltwater interface beneath the freshwater lens fed by land infiltration. As saltwater has a density of 1·025, the freshwater lens floats on it like an iceberg and the inverted cone in the interface is 40 times higher than the matching cone of depression is deep.

19 Ground Investigation

Ground investigation assesses ground conditions prior to starting a construction project.
Site investigation includes legal and environmental aspects, in addition to the ground investigation.

Objectives of a ground investigation vary with the size and nature of the proposed engineering works, but usually include one or more of:

- Suitability of the site for the proposed project;
- Site conditions and ground properties;
- Potential ground difficulties and/or instabilities;
- Ground data to permit design of the structures.

Planning of the investigation then has to be directed towards ascertaining data on three different aspects of the ground conditions:

- Drift and soil conditions, which, especially in the case of cohesive clay soils, involves laboratory tests and application of soil mechanics techniques;
- Rockhead, whose depth is commonly significant to both excavations and foundations;
- Bedrock, whose strength properties and structural variations and likelihood of containing buried cavities are all relevant.

COSTS OF GROUND INVESTIGATION

The extent and cost of ground investigations vary enormously depending on the nature of the project and the local complexity and/or difficulties of the ground conditions.
Expressed as percentages of project costs, the tabulated guideline figures illustrate the contrast between project types but cannot show the contrasts due to differing ground conditions.

Typical Ground Investigation Costs

Project	% Total costs	% Foundations costs
Buildings	0·05–0·2	0·5–2
Roads	0·2–1·5	1–5
Dams	1–3	1–5

The principle of any ground investigation has to be that it is continued until the ground conditions are known and understood well enough for the civil engineering work to proceed safely.
This principle can and should be applied almost regardless of cost – even a doubling of the site investigation budget will generally add < 1% to the project cost – but after an inadequate ground investigation, unforeseen ground conditions can, and frequently do, raise project costs by 10% or more.
Some recent statistics from Britain clearly demonstrate the importance of adequate ground investigation:

- One third of construction projects are delayed by ground problems.
- Unforeseen ground conditions are the main cause of piling claims.
- Half of over-tender costs on road projects are due to inadequate ground investigation or poor interpretation of the data

Savings on the ground investigation budget generally prove to be false economies.

You pay for a ground investigation whether you have one or not

SEQUENCE OF STAGES

Initial stage
- Desk study of available data
- Site visit and visual assessment
- Preliminary report and fieldwork plan

Main stage
- Fieldwork
  - Geological mapping if necessary
  - Geophysical survey if appropriate
  - Trial pits, trenches and boreholes
- Laboratory testing, mainly of soils
- Final report

Review stage
- Monitoring during excavation and construction

These stages are in order of ascending cost so they should form the time sequence to be cost-effective.
It is essential to start with the desk study. As a bare minimum, this is the examination and interpretation of published geological maps, and it is a basis for planning all further investigation.
Any tendency to start an investigation with boreholes is both inefficient and uneconomic. Inefficient because it is often very difficult to interpret borehole logs without the context of some knowledge of the local geology as broadly interpreted from a desk study. Uneconomic because the boreholes may only yield data already available and cannot address any ground problems that should have been identified by a desk study.

DIFFICULT GROUND CONDITIONS

An efficient ground investigation recognizes, during the initial desk study, the possibilities or probabilities of any specific difficult ground conditions occurring within the project site.
It then directs the fieldwork exploration to either eliminate the considered possibilities or determine the extent of the ground difficulties.
The most common difficult ground conditions are:

- Soft and variable drift materials;
- Weathered, weak or fractured bedrock;
- Natural or artificial cavities within bedrock;
- Active or potential slope failure and landsliding;
- Compressive landfill with or without soft spots;
- Flowing groundwater or methane gas;
- Unexpected old building foundations.

UNFORESEEN GROUND CONDITIONS

Construction of a multi-storey car park in Plymouth provided a good example of a project delay due to unforeseen ground conditions.
The site extended over 200 × 70 m, with a complete layer of drift and therefore no bedrock exposure.
15 boreholes found rockhead at 5–10 m deep.
Piling work then found a deep rockhead gully with steep sides; this extended across nearly 10% of the site; all the boreholes had missed it.
Project was delayed, while 100 probes were used to further explore rockhead.
Bored piles were needed over sloping rockhead.
The gully had been formed by solution of a narrow unmapped limestone band; it had been impossible to foresee.
In this case, the need for more exploration probes was only apparent with the benefit of hindsight.

WALKOVER SURVEY

An early site visit combines with the desk study to recognize any possibilities of difficult ground conditions, so that a planned field exploration is cost-effective.

Check-list of aspects requiring only observation:

Correlate ground features with geological map; vegetation may relate to rock type.
Local exposures: check stream banks, road cuts and quarries for geological details and soil profiles.
Land use: signs may remain of past use for mining, old tips, backfill, quarries, buildings, basements.
Physical features of ground may be interpreted; escarpments, moraines, terraces, floodplains, peat flats easily recognized.
Breaks of slope: all must have a reason: edge of erosion profile, geological boundary, or artificial.
Lumpy ground: created by hollows or hummocks or a combination of both. May be caused by any of a variety of conditions: sinkholes, crown holes, mine subsidence, mineshafts and waste heaps, quarry waste, moraine, landslip, solifluction. All except moraine provide potential appropriate engineering hazards and require investigation.
Existing structures: check for distress in buildings and stability of old cut slopes.
Landslip: disturbed ground, displaced or damaged structures, deformed trees.
Groundwater: sinkholes, springs, seepages, solutional features, stream levels, flood potential.

PHOTOGEOLOGY

Geological interpretation of air photographs can be a valuable part of the desk study of some sites.
Interpret from vertical air photographs, scale normally about 1:10 000 on contact prints.
Photographs taken with 60% overlap along flightpath, so view through stereoscope to see 3-D image with vertical exaggeration of relief.
Use in ground investigtion to identify local contrasts, anomalies and relief features visible on photos and which relate to ground conditions.
Black and white photos: widely available and usually most cost effective.
Colour photos: expensive, and colour may disguise some features.
Infra-red photos: sensitive to temperature, so useful to trace emerging groundwater from seepages and small springs.
Multi-spectral images: not widely available at large scale, and need specialist interpretation. All photos reveal little of the geology in urban areas and beneath thick tree cover.

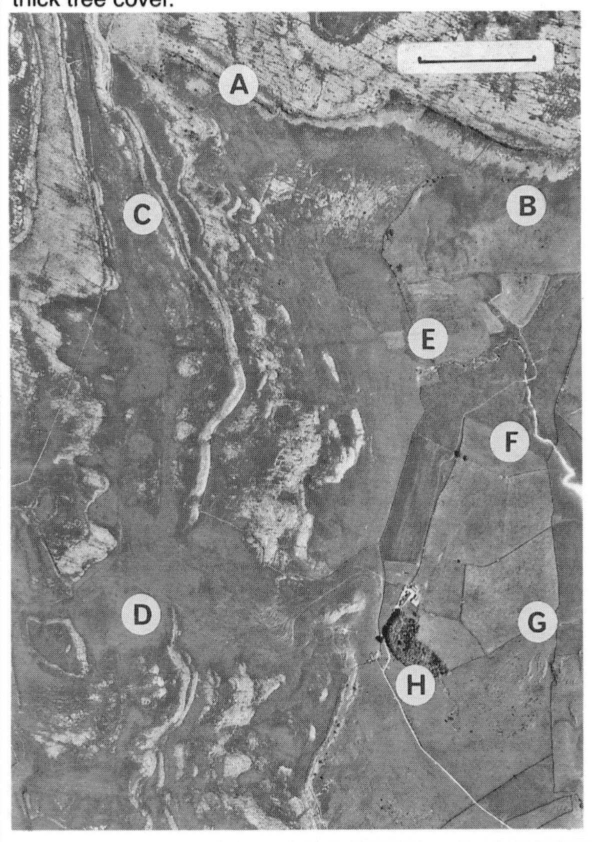

Interpretation of geology

Photographs show vegetation and soil; generally these are related to drift and bedrock therefore contrasts on photos can be interpreted as contrasts in ground conditions.
Interpretation is largely based on three factors:
Tone: generally related to water content of soil and plants; dark = wet clay; light = dry sand.
Texture: includes drainage channel density and patterns, rock banding and lineations, and patchy or mottled ground.
Trend: single linear features or correlated anomalies, may trace geological boundaries or structures.
Geomorphological features can be directly identified, notably landslips, moraines, sinkholes, old channels, terraces, breaks of slope, etc.
Distinguish man-made features by association, e.g. ploughing patterns relate to field boundaries.
Land use may relate to rock or soil type.

Ground types on photographs

Alluvium: light sand and dark clay tones, level ground, channels, meanders.
Glacial till: irregular relief, hummocks and drumlins, patchy tones.
Landslips: arcuate head scars, concave slopes, debris hummocks, lobate toes.
Bedrock: angular jointed textures in stronger rocks; most bare rock has lighter tone, notably white limestone.
Rock permeability: relate to drainage channel density: many = clay; few = sandstone; none = limestone.
Structures: may show by linear trends, tonal zones.
Faults: straight features, may disappear under drift.
Old shafts, active or filled sinkholes: spot anomalies with dark or light tonal contrast.

Air photograph from the Yorkshire Dales

North is top left; scale bar is approx. 300 metres. Relief cannot be seen without stereopair; valley at lower right has high ground to both NW and NE.
A: bare limestone pavement with clear jointing.
B: bare screes and grassed lower slopes.
C: scars along contours in horizontal limestone.
D: tongue of glacial till buries limestone scar.
E: spring from limestone feeds stream flowing over impermeable rock.
F: valley floor in soft slates with drift and soil cover.
G: bands of greywacke form small outcrop scars.
H: tree plantation, farm and gravel road.

20 Desk Study

Every site investigation should begin with a desk study. This is an office-based exercise (therefore inexpensive) in gathering published and available information.
An initial site visit is made during the desk study; ground conditions are more readily visualized once the site has been seen, and potential ground problems can better be appreciated (section 19).
Sources of data vary with type, size and location of site but generally fall into five groups, as below.
They also vary between countries: Britain has a greater variety of historical and recorded data than many and provides a comprehensive example of the desk study material which can be sought.

GEOLOGICAL MAPS AND RECORDS

British Geological Survey (BGS) has published and unpublished data which is readily available.
Published geological maps are normally the first desk study target; for a small investigation on uncomplicated ground they may provide all the data required.
Printed material available from the BGS includes:

- Map and book series tabulated opposite.
- Reports on applied geology related to planning and development, with multiple derived maps and engineering data summaries; available for 50 urban areas, as valuable guides to difficult ground.

Digital data is now the main BGS output, over the web for licensed regular users, or to single requests.
DigMap has digitized map for the whole of Britain at scales of 1:10 000 or 1: 50 000 data always updated (so better than paper copies), available on demand as digital file or print-out with appropriate descriptive text. Layers of data include geology, drift thickness, landslip, made ground, and outline geohazard potential.
Geoscience Data Index at www.bgs.ac.uk/geodata has outline geological and surface maps to locate sites of borehole records, maps and reports coverage and lists of all available data, with an ordering facility.
Enquiries@bgs.ac.uk will produce on request:-

- Maps, reports, borehole logs, mining records, site investigation reports and laboratory test data held by the National Geological Records Centre.
- Reports on specified areas, sites or geohazards based on interpretations by a geologist; these are consultancy reports produced at commercial rates.

National surveys on mining instability (1992) and on natural cavities (1994) were produced for DoE and are available in hard copy; useful guides to potential hazards in unfamiliar areas. Data from the landslide survey (1988) is now incorporated in BGS database.
BGS library and borehole core store, near Nottingham, are accessible at nominal charges for commercial use.

LOCAL SOURCES

Much detail on ground conditions from local residents, farmers, historians, societies, universities and council authorities. Difficult to trace without local knowledge, often for diminishing returns, but can yield useful pointers.

Data Sources in USA

US Geological Survey (USGS) publishes a huge list of geological maps, reports, topographic maps, air photographs and digital data; accessible through www.usgs.gov or at the offices of State Geological Surveys (which also have their own publications).

MINING RECORDS

Coal mining has left large areas of undermined, potentially unstable ground in Britain and in many other countries.
Coal Authority is obliged by law to keep and provide data on all aspects of coal mining in Britain. Mansfield office, at 01623 427162, or check www.coal.gov.uk
Mining reports for engineering enquiries cost £34, and notify known past, present and future underground and opencast mining, recorded shafts and adits, and claims for subsidence damage since 1984.
Current mining: recorded on seam plans at 1:10 000; these show recent mining and directions of working, and are available for inspection on request.
Subsidence predictions are not provided, but advice may be gained from a consultant or the mine operator.
Past mining is incompletely recorded; records are only complete since 1947, and seam maps before then are generalized and incomplete. As old records are unreliable, assume all workable seams (notably < 100 m deep, above water table, > 0·6 m thick) have been worked, unless proved otherwise (normally by boreholes).
Abandonment plans, required by law when a mine closes, have often been lost; many with Coal Authority at Mansfield, available for inspection; can give useful detail, but may be difficult to relate to present locations.
Shaft register records all known shafts in coalfields, on 1:2500 maps with files noting depth, size, capping and treatment if known. Sites are noted in mining report and extracts of full data are available at £11 per shaft.
Opencast mining sites, and the backfilled areas, are recorded on 1:10 000 maps.
Mining other than coal has no controlling authority, so records on old stone and metal mines are extremely erratic in coverage and reliability, and can be difficult to trace. Prime sources are county authorities (but BGS in Scotland); some with many mines (inc Derbyshire and Cornwall) have systematic records and search facilities; others have little or no data submerged in archives.
Records from a nationwide search after the Lofthouse mine disaster in 1972 (due to forgotten workings) is now held by the Coal Authority or the counties.

The lost shafts of Wigan. In 1958, 500 old mine shafts were known within boundaries of Lancashire town of Wigan. In 1980, after years of redevelopment and site clearance, 1700 shafts were recorded in the same area. How many more shafts remain unmapped?

TOPOGRAPHICAL MAPS

Old maps show features no longer visible on the ground and therefore omitted from later maps. Best are first edition ordnance surveys (6" = 1 mile) of about 1870; mostly in local libraries for reference.
Simple comparisons with new maps may show:

- Old quarries, mines, buildings, past land use;
- Old streams, ponds, valleys lost due to landfill;
- Erosion changes in rivers, coastlines and landslips.

AIR PHOTOGRAPHS

Black-and-white photographs, around 1:10 000, widely available, about £16 per print, covering 2 km square.
Useful for site detail and photogeological interpretation in certain conditions (section 19).
Source: National Air Photograph Library at Swindon, www.english-heritage.org.uk/knowledge/nmr

BRITISH GEOLOGICAL SURVEY – MAIN PUBLICATIONS

	Coverage and availability	Information provided
Maps		
1:50 000	330 sheets for England and Wales; most now available as paper copies; complete coverage in digital form, available as extracts on demand.	Separate solid and drift, or combined, editions; outcrop boundaries and dips only; no underground data; with brief descriptions of rock and drift types; landslip, cambering and other features on newer maps.
1:25 000	45 sheets only, of areas of geological or local interest, inc some new towns.	Information as on the 1:50 000 maps, with benefit of larger scale.
1:10 000	75% of Britain is available on demand as digital file or print-out; paper sheets on sale while stocks last.	Solid and drift geology, shown with minimal colour; some underground data and borehole depths; show landslips, made ground, backfill, and some shafts.
Texts		
Regional Guides	20 guides cover the whole of Britain, each with 100–150 pages.	Broad overviews with rock descriptions keyed to geological history; useful where 1:50 000 maps do not have enough rock description; minimal data on drift materials, and none on engineering geology.
Sheet Explanations	32 pages for each 1:50 000 map, except where old larger memoir in print.	Valuable illustrated reviews to help interpretation of map data; only available for recently published maps.
Technical Reports	Sheet descriptions are 48 or more pages, each for one or more 1:10 000 maps sheet, or a 1:50 000 map; most only printed on demand, or data is supplied specific to a map extract include old 1:50 000 sheet memoirs.	Comprehensive reviews of mapped and recorded geology; masses of descriptive geological detail, keyed stratigraphically; chapters on drift sediments and mining are usually short; later editions have useful summaries of hydrogeology, geological hazards, slope stability and made ground, but lack the detail in the urban geology planning reports.

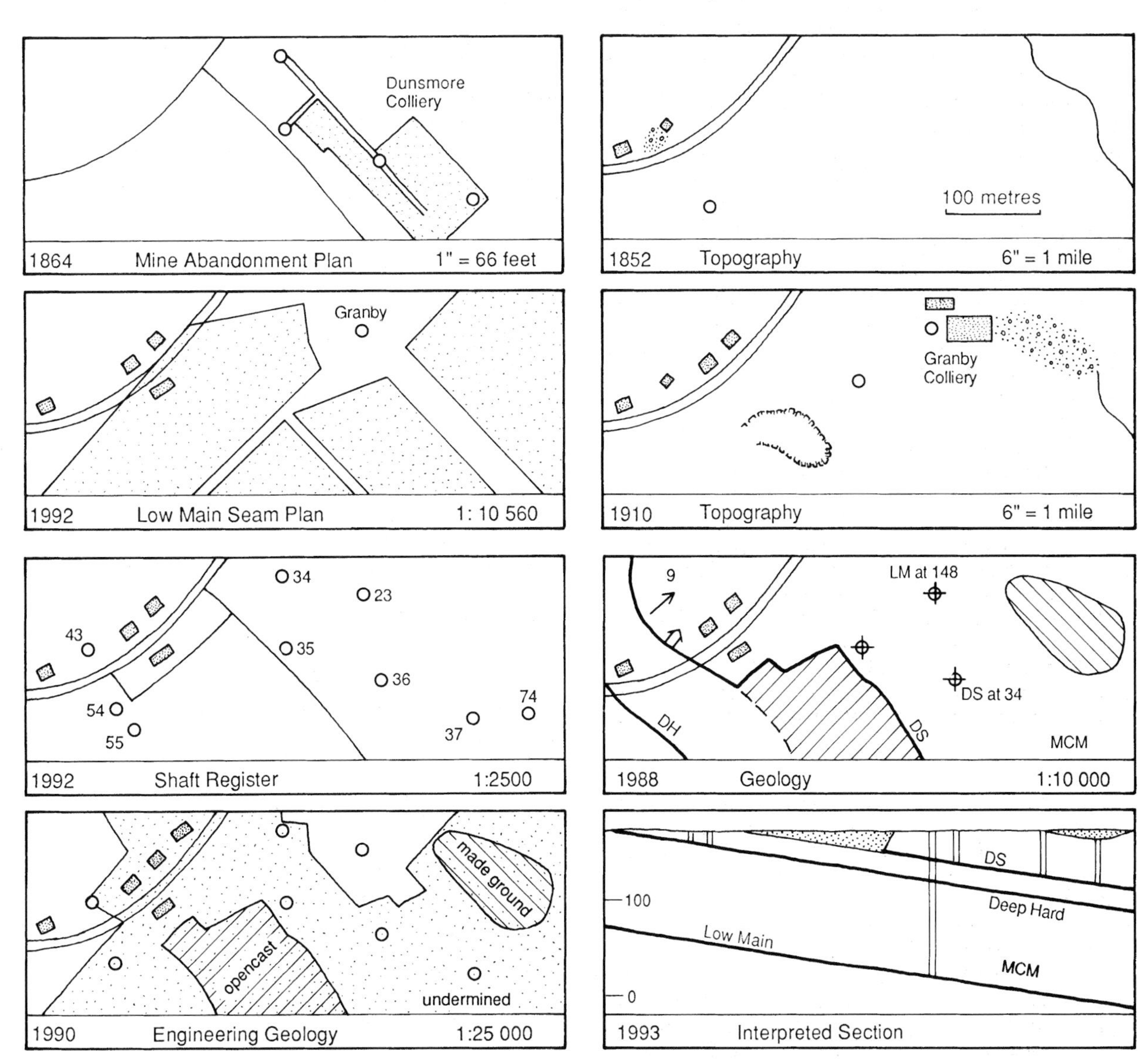

Desk study information on a level site with difficult ground conditions within a Midland coalfield.
Extracts from available old and modern maps, each redrawn to cover the same area, from originals at stated scales.

21 Ground Investigation Boreholes

BOREHOLE DRILLING METHODS

Holes may be drilled with a great variety of commercially available rigs; methods fall mainly into three groups, dictated by the need for soil or rock penetration and required sample or core recovery.

LIGHT PERCUSSION DRILLING

Mobile A-frame, easily erected, with power winch.
Steel shell is driven into ground by weight repeatedly dropped 1–2 m and lifted by cable over A-frame (hence 'cable and tool' rig).
Only for shallow exploration of soils and soft clay rocks.
In clays, smooth shell is driven by weight dropped onto it and soil adheres inside.
In sands, whole weighted shell may be surged and dropped, and soil is held inside by hinged clack valve.
Can add light rotary drive for auger drill in clays (hence 'shell and auger' rig) but less suitable for most site investigations as sample is disturbed.
May use chisel head for limited rock penetration.
Widely used, as all sites need soil investigation; usually with 100 mm sampler inside 150 mm casing, reaching depths of 15–40 m.

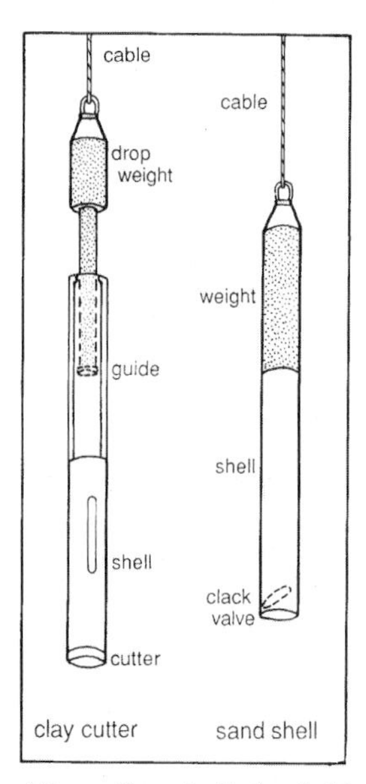

Alternative shells for light percussion drills

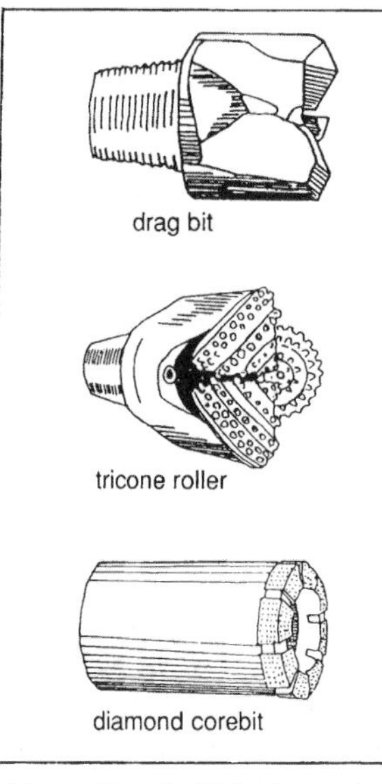

Alternative drill bits for rock penetration

TRIAL PITS and TRENCHES

Cheapest method of shallow soil exploration.
Dug with any site excavator with backhoe.
Usually 2–5 m deep; may need temporary support or safety cage to allow full inspection of exposed walls.
Especially useful in variable man-made fills.
Valuable in disturbed or slipped material, including soliflucted head, as shear surfaces may be recognized in clean cut walls and not in borehole cores.
Can cut block samples, or drive in U100 sample tubes with backhoe, or test load a plate on pit floor.
Trenches can expose rockhead in search for fractures or outcrops.
Avoid trenches precisely on foundation sites; backfill with compacted soil or lean concrete.
Pits and boreholes may need sealing to prevent groundwater movement through breached aquiclude.

Light percussion rig sampling soils to a depth of 10 m for a construction project in Derbyshire

ROTARY CORING

Truck-mounted rig with full rotary drive which can be applied with downward force.
Air, water or bentonite mud flush, pumped down inside drill string, and washes chippings back up outside.
Tip of cylindrical drill bit is tugsten carbide or with diamond inserts (hence 'diamond drilling'). Normally double tube barrel holds full core in inner non-rotating barrel about 1–3 m long.
Can penetrate any soil or rock to >100 m deep.
Commonly use N size, providing 54 mm diameter core from cased hole of 89 mm diameter. Larger diameter is better in weaker rocks.
Continuous flight hollow-stem auger can be used in clays; not common in exploration.

ROCK PROBING

Rotary percussion rig with hammer action capable of rock penetration.
Tricone roller or drag bits with air or water flush to remove chippings (hence 'open hole drilling').
Truck mounted to provide rotation and downward force; large rigs can reach >100 m deep.
No core recovery (hence 'destructive drilling') but much cheaper than diamond drilling.
Penetration rate indicates strength of rock, soil or voids; flushed chippings can be examined; flush loss also indicates cavities.
Wash boring uses water flushing in driven shell to probe soils (common in USA but rare in Britain).
Hand held pneumatic drill can reach 8 m in uniform rock.
Used mainly to locate cavities in rock and rockhead beneath soils.

BOREHOLE RECORDS

All boreholes must be logged as completely as possible to be cost-effective.
Best to use conventional symbols for ease of reading.
Log must record at least the data on this example, on some style of conventional prepared booking sheet, though there is no single all-purpose format.
Description, thickness, depth, and the scale pictorial log provide a basic understanding.
In situ tests quantify ground properties.

Standard Penetration Test is easiest borehole strength test for soils; N value increases with strength (section 26). Or use Cone Penetration Test (section 26).
Point Load Strength is field test on borehole core of rock (section 24).
Rock Quality Designation is measure of fracture density in rock (section 25), and core recovery is measure of weak broken zones; both values increase with quality and integrity of rock mass.

Water table and inflow points should be recorded, and permeability may be determined by packer tests (section 18).
Sample points are noted.

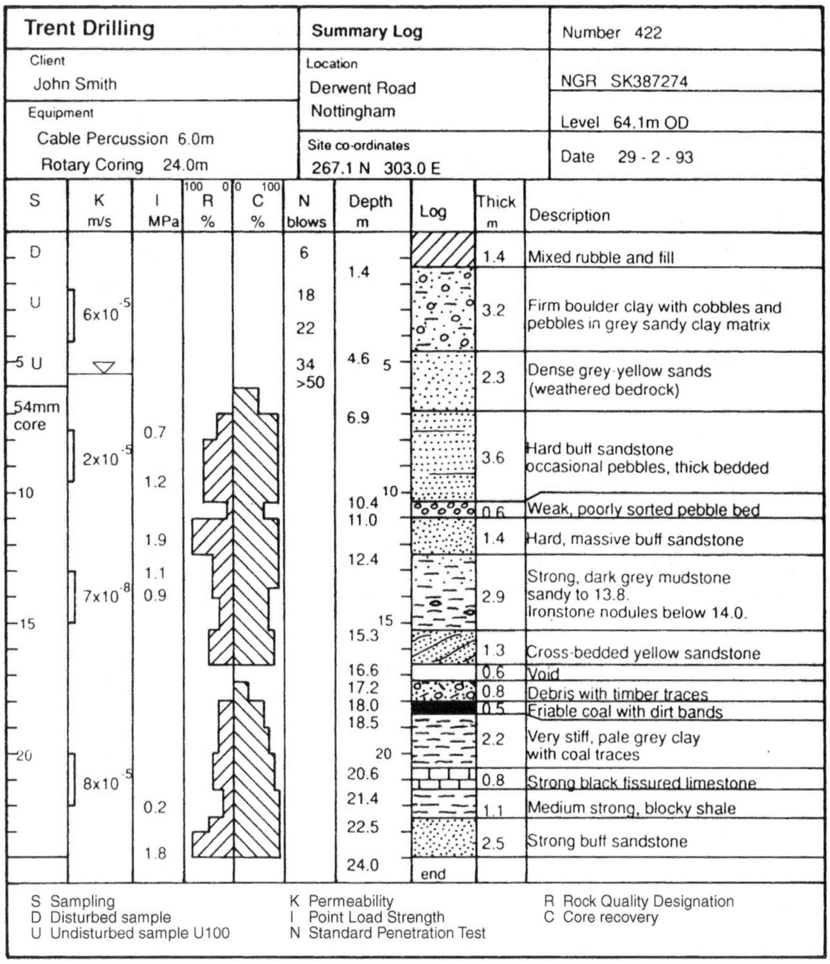

Trent Drilling
Summary Log
Number 422
Client
John Smith
Location
Derwent Road
Nottingham
NGR SK387274
Equipment
Cable Percussion 6.0m
Rotary Coring 24.0m
Site co-ordinates
267.1 N 303.0 E
Level 64.1m OD
Date 29 - 2 - 93
S
K m/s
I MPa
R %
C %
N blows
Depth m
Log
Thick m
Description
D
U
5 U
54mm core
10
15
20
6x10^-5
2x10^-5
7x10^-8
8x10^-5
0.7
1.2
1.9
1.1
0.9
0.2
1.8
6
18
22
34
>50
1.4
4.6
6.9
10.4
11.0
12.4
15.3
16.6
17.2
18.0
18.5
20.6
21.4
22.5
24.0
end
1.4 Mixed rubble and fill
3.2 Firm boulder clay with cobbles and pebbles in grey sandy clay matrix
2.3 Dense grey-yellow sands (weathered bedrock)
3.6 Hard buff sandstone occasional pebbles, thick bedded
0.6 Weak, poorly sorted pebble bed
1.4 Hard, massive buff sandstone
2.9 Strong, dark grey mudstone sandy to 13.8. Ironstone nodules below 14.0.
1.3 Cross-bedded yellow sandstone
0.6 Void
0.8 Debris with timber traces
0.5 Friable coal with dirt bands
2.2 Very stiff, pale grey clay with coal traces
0.8 Strong black fissured limestone
1.1 Medium strong, blocky shale
2.5 Strong buff sandstone
S Sampling
D Disturbed sample
U Undisturbed sample U100
K Permeability
I Point Load Strength
N Standard Penetration Test
R Rock Quality Designation
C Core recovery

Rotary rig coring drift and rock to a depth of 30 m for a road bridge in North Wales

HOW MANY BOREHOLES, HOW DEEP?

Spacing: buildings 10–30 m apart;
road lines 30–300 m apart;
landslides at least 5 in line for profile.

Depth: 1·5 × foundation width, below founding depth,
plus at least one deeper control hole to 10 m below foundation unless rockhead found;
3 m below rockhead to prove sound rock;
probes to 3–10 m to locate rock cavities.

These are only rough guidelines.
Spacing and depth may be varied considerably in light of local conditions and appropriate to size of structure.
Cavernous rock may need probes at each column base.
Old mine working may need proving to depths of 30 m, and location of old shafts may need probes on 1 m grid.

BOREHOLE COSTS

Drilling costs are best estimated as accumulation of:

- Cost to supply rig onto site
- Cost of set-up on each new hole
- Cost per metre of hole drilled

Table shows approximate relative costs (2001 figures in £)

	on site	+ per hole	+ per m
Light percussion, soil <10 m deep	300	30	14
>10 m deep	300	30	16
Probing in rock or soil	350	40	12
Rotary coring in rock	350	40	40
Trial pits, 4 m deep, backfilled	250 for 4 pits		

Costs vary with number of in situ tests required.
Probes are cheaper on close-spaced grid and in uniform rocks which do not require casing of hole.

22 Geophysical Surveys

The techniques of geophysical exploration involve the remote sensing of some physical property of the ground using instruments which in most cases remain on the ground surface.
Passive methods accurately measure earth properties and search for minute anomalies (local distortions within the overall pattern). These include gravity and magnetic surveys (and radioactivity which is inapplicable to site investigation).
Induction methods send a signal into the ground and pick it up again nearby. These include seismic, electrical, electromagnetic and radar surveys.

Interpretation of geophysical surveys invariably requires some borehole data, either to calibrate profiles or to test-drill anomalies.
Geophysical surveys have two main uses in ground investigations:

- Filling in detail between boreholes;
- Searching a large area for anomalies before drilling.

Geophysics is low cost compared to multiple boreholes. It can be cost-effective in site investigation in certain difficult ground conditions where a particular type of geophysical survey may be appropriate; there is no single geophysical system applicable to all problems.

GROUND PROBING RADAR (GPR)

Trolley-mounted transmitter and receiver record microwave electromagnetic radar signals reflected from ground contrasts. High cost equipment, needs trained operator and assistant.
Ground cross-section is produced as computer output; some outputs are complicated by reflection interference, but many are realistic displays. Calibrate depth and materials with borehole.
Limited depth penetration is main restriction: 10–20 m in dry sand, only 1–3 m in wet clay.
Can tow behind car at 5 km/h for continuous profile.
Can use to map shallow drift profiles, filled sinkholes, shallow voids.

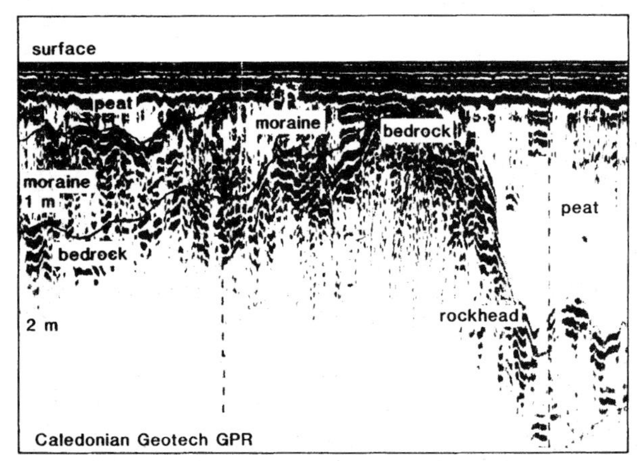

Ground radar profile of drift and rockhead for a road project in Scotland. Transverse length is 200 m

ELECTRICAL SURVEYS

Numerous methods applied successfully to mineral exploration.
Resistivity surveys with Wenner arrays of four ground electrodes can be used to map lateral and vertical changes in ground conditions.
Difficult to interpret; limited use in site investigation.

SEISMIC SURVEYS

Shock waves, produced by hammer-blows, explosions, etc., are reflected or refracted on geological boundaries.

Reflection seismic
Seismic waves reflected from deep strata boundaries.
Successfully used for all primary oil exploration.
Difficult and little used in shallow ground investigation.

Refraction seismic
Seismic waves refracted at shallow geological boundaries and returned to surface.
Drop-hammer or 3 kg sledge hammer adequate for 20 m penetration; deeper with explosive shock source; small geophones detect wave arrivals; low cost equipment, 2 man operation.
Refraction relies on faster layer at depth: rockhead is ideal boundary to detect, with slow soil over fast rock.
Graphical plot of first wave arrivals reveals both velocities and boundary depth. Other simple relationships apply to 3-layer situations and dipping or stepped boundaries.

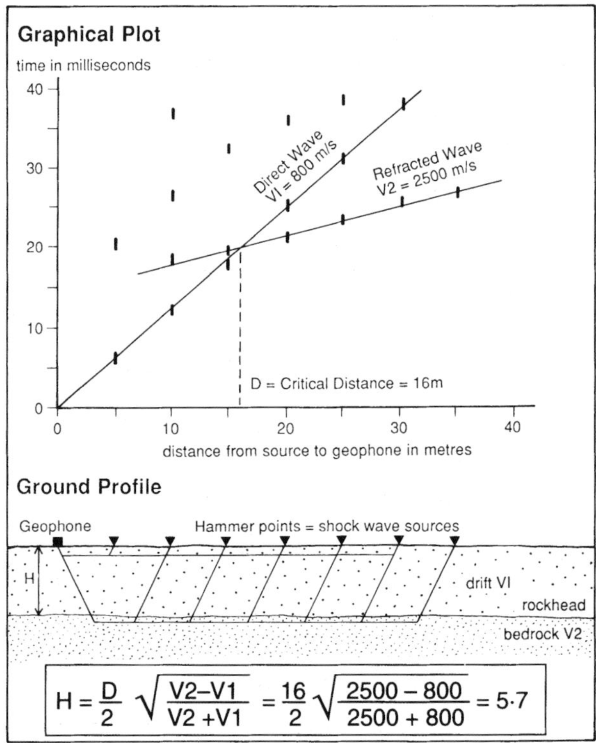

$$H = \frac{D}{2}\sqrt{\frac{V2-V1}{V2+V1}} = \frac{16}{2}\sqrt{\frac{2500-800}{2500+800}} = 5{\cdot}7$$

Seismic profile over alluvium above mudstone

Cross Hole Seismic
Seismic waves transmitted between boreholes have scope for detection of isolated voids and complete tomographic profiling, but require trained operators.

Seismic velocities (speed of shock waves through rock) increase with strength of rock, and decrease with more fracturing (related to RQD, see section 25).

Typical seismic velocities (Vp)

Drift and soil	500–1500 (m/s)
Shale and sandstone	1500–4000
Limestone	3000–5000
Granite	4500–5500
Fractured rock	unfractured Vp × RQD/100

MAGNETIC SURVEYS

Record distortions of Earth's magnetic field.
Proton magnetometer measures total field; low cost, robust equipment. Measures to 1 nanoTeslar (1nT = 1γ = about 1/50 000 Earth's field).
Simple to use,10 seconds per station, 1 man operation.

Dipole anomalies – positive next to negative, so easily recognized – are due to vertical linear features, e.g. buried mine shafts. Unlined shafts with fill which is magnetically similar to wallrock may go undetected.
Fences, drains, powerlines, iron-rich fill prohibit use.

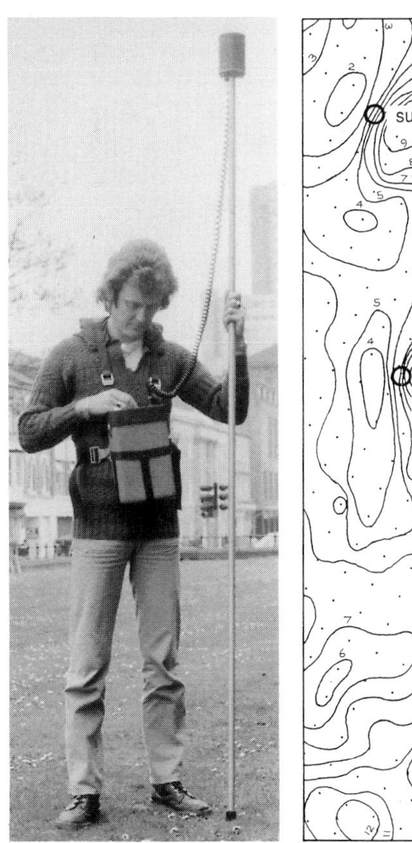

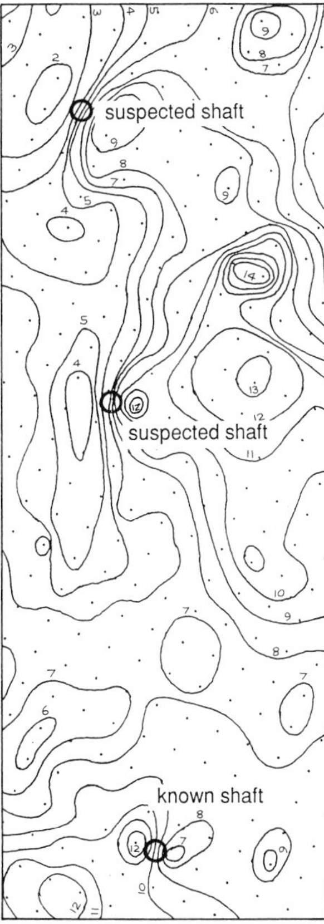

Proton magnetometer and map of a magnetic survey on a site in the Derbyshire coalfield. Stations are on 3 m grid; contours are at 100 nT intervals above a base of 48 000 nT. Dipole anomalies lie over one known shaft and indicate the location of two more shafts

USES IN GROUND INVESTIGATION

The new geophysical techniques can be, and have been, applied effectively to certain specific problems in ground investigation:

- Search for unknown cavities: GPR if depth < 10 m, or gravity survey if depth < size
- Search for suspected mineshafts: magnetic survey.
- Trace lateral contrasts, notably between sand and clay, in shallow drift: GPR, electromagnetic survey.
- Rockhead profiling between boreholes: refraction seismic survey.
- Estimate rock fracturing ahead of new tunnel drive – seismic survey, reflection or refraction depending on depth of cover.

Magnetic searches for buried mineshafts are simple enough for operation and interpretation by untrained personnel with low-cost rented equipment.
All other geophysical surveys are best interpreted by specialists working as part of a ground investigation team.

GRAVITY SURVEYS

Record minute variations in Earth's gravitational force.
Gravimeter measures length of internal weighted spring; high cost, delicate instrument. Measures to 0·01 gravitational unit (1 gu = 10^{-6} m s^{-2} = 0·1 mgal).
Ten minutes per station, one man operation.

Negative anomalies due to underground voids (cave or mine) or low density soil or rock (in buried valleys or sinkholes); both significant to engineering.
The limit is set by background noise, but microgravity surveys with computer analysis of closely spaced data points can recognize voids with diameter much less than their cover depth. Can trace small mines to depths of 20 m, and larger limestone features to much deeper.
Depth and size of void may be interpreted from shape of anomaly, but normally drill all negative anomalies.

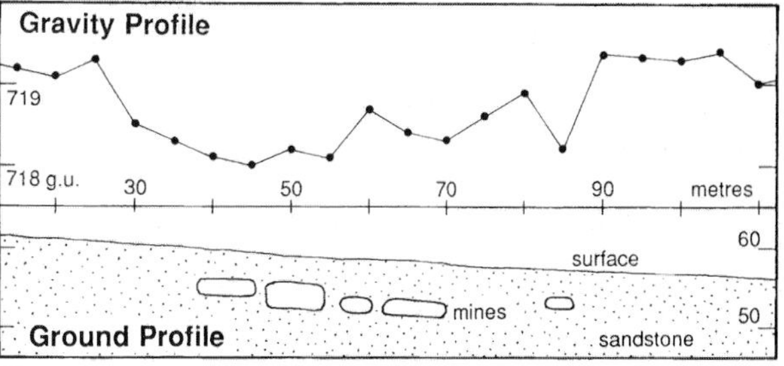

Gravity surveys over sand mines in Nottingham

ELECTROMAGNETIC SURVEYS

Non-contacting terraine conductivity meter creates electromagnetic field in ground and measures field intensity 3·7 m away; low cost equipment, simple to use, similar to large metal detector. Measures mean conductivity in hemisphere of ground reaching 6 m deep (deeper on some meters).
Continuous reading, 1 man operation.
High conductivity of clay, basalt and water, contrasts low conductivity of sand and limestone.
Can use to map shallow lateral changes: clay-filled fissure zones, filled sinkholes, rockhead steps, alluvial channel fills, high permeability fracture zones.

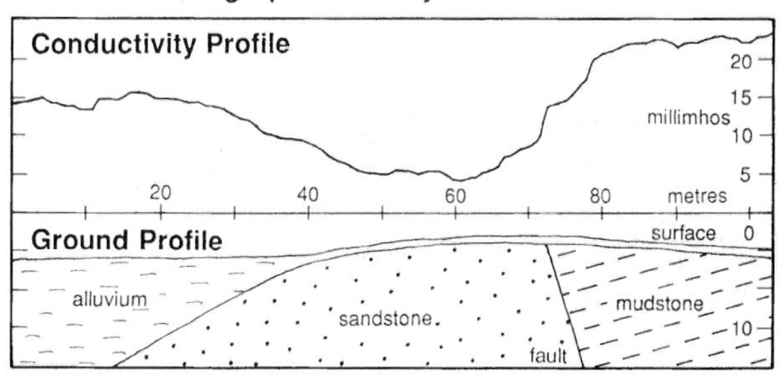

Electromagnetic traverse over faulted mudstone and sandstone with variable alluvium cover

GEOPHYSICAL SURVEY COSTS

Comparisons of costs are tenuous because each method is best applied to only certain ground problems. Rough guide is given by approximate coverage which can be achieved for a given fee – in this case £500 at 2001 prices.

Microgravity survey	0·1 ha on 2 m grid
Magnetic survey	0·5 ha on 3 m grid
Electromagnetic survey	0·5 ha on 3 m grid
Ground probing radar	600 m of line profile
Seismic refraction	5 soundings to 20 m deep
Borehole	1 cored hole 10 m deep

23 Assessment of Difficult Ground

Each ground investigation must be tailored to the local potential problems

Overview of possible hazards recognizes the geological conditions which determine the scale of each: for any one site, many potential hazards can be eliminated in the desk study phase, while others demand specific investigation techniques.

GROUND SUBSIDENCE

Beyond the acceptable limits of structural settlement on soils, this can only occur on certain rock types.

Limestone with solution cavities provides the most difficult ground (section 29).

Can also occur on clay, peat, loess, chalk, salt, gypsum, basalt (sections 27, 28).

Major potential hazard over any mined ground.

SOFT GROUND

Mostly provided by alluvial clays, lake sediments, organic soils, young clays, artificial made ground.

Laboratory testing to find bearing capacity.

CAVITY SEARCH

Natural and artificial cavities in rock are notoriously unpredictable in their locations.

Local building codes may require probes for 5 m at each column base, with central vertical hole and holes splayed out at 30° on each corner.

ROCKHEAD RELIEF

May influence foundations and tunnelling.

Buried valleys are most likely in areas of glaciation and meltwater erosion; larger features may be traced by seismic survey.

Solution features and pinnacles dissect limestone rockhead; may need many boreholes or soil stripping.

Tunnelling up through rockhead is major hazard; rely on boreholes from surface and probing in advance of heading. Latter would have averted Lötschberg tunnel disaster (section 15).

SINKHOLES

Notable hazard in soil or drift cover on cavernous or fissured limestone (section 27).

May need grid of probes to rockhead; case wash borings to prevent flushing and sinkhole inducement.

SPT values can sometimes give an indication of incipient collapse of soils into an enlarging void beneath:

N = 5–10 for normal soils;

N = 0–2 for soils in tension and about to fail.

SLOPE FAILURE

Potential threat of landslides depends on ground slope, rock type and inclinations of rock structures; limit angle for stable slopes may be around 10° in clays, 30–40° in well fractured rocks, and up to vertical in strong, massive rocks (section 37).

Local data, including water conditions, allows hazard zoning based on these factors.

Trenches are useful to assess shallow slide geometry, and are the most reliable method for recognizing solifluction shears – they were not used prior to the Carsington Dam failure (section 16).

Monitor potentially active slides through a wet season.

EARTHQUAKES

Destruction can be minimized by appropriate design.

Active fault zones recognized by displacement of recent sediments; mapping permits some authorities to limit new buildings within 15 m of known faults (35 m for larger structures), with extra 15 m setback on faults not accurately traced.

Geological mapping constructs 4 zones of ground with respect to vibration amplification in sediments: recent muds (most unstable); thick drift cover; thin drift cover; bedrock outcrop (relatively stable).

Liquefaction hazard mapping records well graded, low density soils with high water table – the least stable.

Full earthquake hazard mapping also includes landslide potential, tsunami threat and dangers from dam failures.

ROCKHEAD PROFILE UNDER MOTORWAY

Site investigation for motorways round Birmingham.

Pre-construction boreholes were spaced 30–150 m along proposed roadline, 5–40 m deep to sample soils and prove rockhead.

Found deep buried valley, offset 300 m from modern river course, filled with silts, sands and soft clays.

Photograph of the viaduct and profile of the ground where the M5 motorway crosses the buried valley revealed by boreholes beneath the River Tame

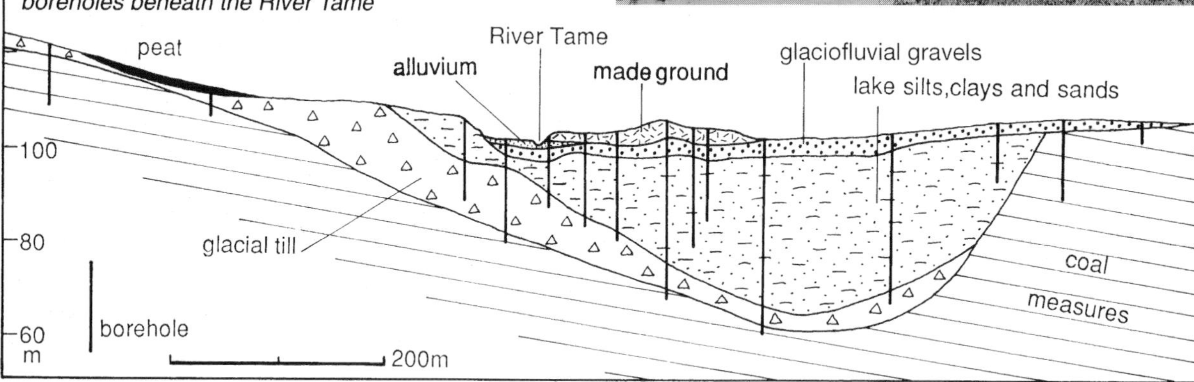

CAVITIES UNDER REMOUCHAMPS VIADUCT

Motorway viaduct in Belgium built on folded limestone, shale and sandstone.
Desk study showed cavernous limestone.
Microgravity survey gave indefinite results.
4–8 cored boreholes on each of 13 pier sites; found strong sandstone, soft weathered shale, and irregular rockhead over solid limestone.
Footings on shale redesigned for low loading.
Excavation found caves beneath 2 of the 5 footings on limestone; had been missed by all the boreholes.
Grid pattern of new probes on all limestone footing sites; 300 holes found no more caves.
First borehole programme inadequate, second excessively cautious.
Caves were filled with concrete; one pier was relocated by 15 m to avoid largest cave.
Ground conditions increased contract cost by 15%, including delay costs when caves were found.

Remouchamps Viaduct

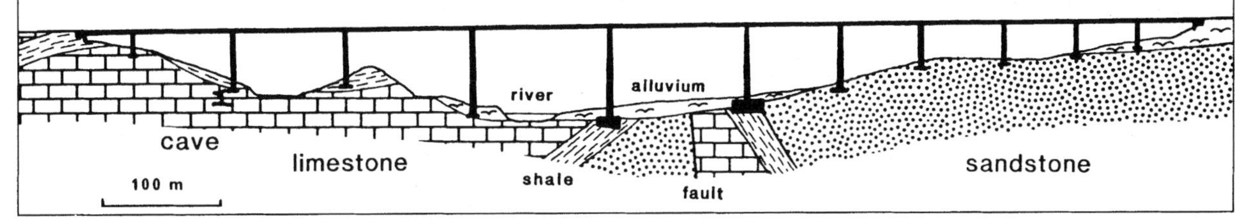

REDEVELOPMENT SITE IN COALFIELD

Site for housing within old Derbyshire coalfield.
Staged ground investigation revealed difficult ground.
Desk study: coal outcrop, old brick pit, shallow old mines, approximate locations of 5 shafts.
Site visit: recent fill on whole site, one sinkhole probably over old shaft.
Boreholes: 4 light percussion holes showed 4–12 m soft fill over weathered shale. 2 trial holes found edge of fill. Methane tests gave low values.
Magnetic survey: 5 dipole anomalies, 3 of them within 5 m of recorded shafts.
Probing: needed on 1 m grid on all shaft locations and magnetic anomalies until shafts are positively located; deep cover of fill made costs prohibitive for the value of the site, so project temporarily abandoned.

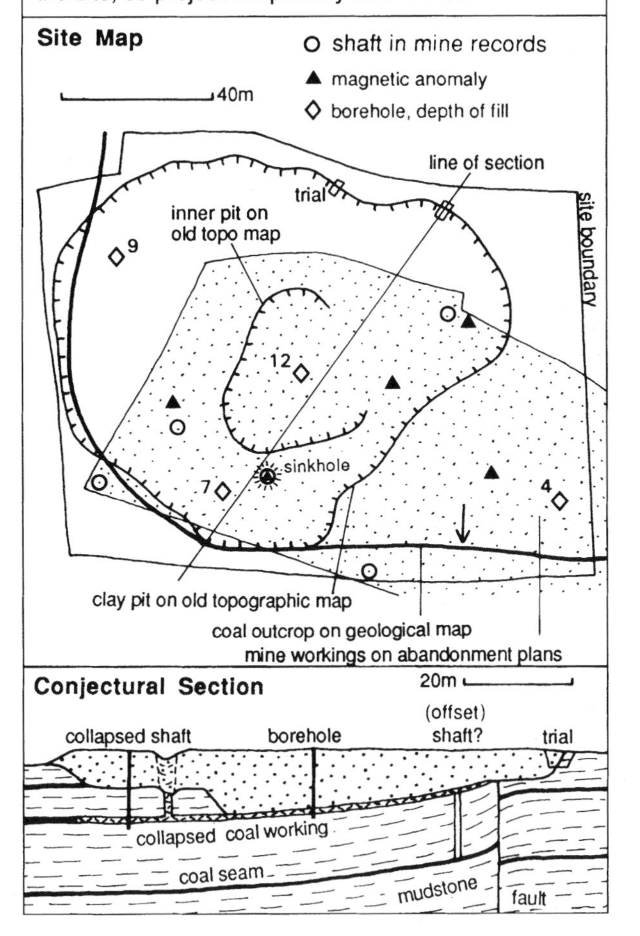

MINED GROUND

Best guide to potential hazard is historical data obtained on desk study (section 20).
Coal measures are the most extensively mined rocks; granites and limestones have the most mineral veins.
The 30 m guideline: mines < 30 m deep are most likely to collapse and endanger the surface; mines > 30 m deep are generally, but not always, stable.
Backfill in opencast mines is usually prone to high compaction; highly variable and best load tested in situ.
Redevelopment of second-hand ground nearly always finds 'unforeseen ground conditions'; areas of mining and fill provide worst case.

HAZARD ZONING OVER OLD COAL MINES

Coal outcrop positions, dip directions and dip amounts are enough to identify zones on geological map where shallow mining could occur.

A
B
30m
30m
coal seam
Profile
Plan
A
B
hazard zone
open cast
17
coal outcrop
30m cover line

BURIED SHAFT SEARCH

Must check every suspected site before construction.
Shaft register of Coal Authority (or local council) generally gives location only within about 10 m.
Check for any disturbance of ground or drainage.
Check old maps and air photos; ask local residents.
If soil cover < 2 m, trench or strip with backhoe.
If cover is thicker, geophysics may be useful; magnetic survey is generally best and cheapest (section 22).
Last resort is expensive probing to rockhead: start at 'best-guess' location, then spiral out on a grid; probe spacing must be 0.5 m less than the suspected shaft diameter, so 1.0–1.5 m for small old shafts, and 2–4 m for larger, more recent shafts.
Many searches have needed more than 50 probes.

24 Rock Strength

Strength of intact rock depends on component mineral strengths and the way they are bound together – by interlocking or cementation.
Rock mass strength applies to a mass of fractured rock within the ground and largely relates to the fracture weaknesses.
Hardness is not directly related to strength; normally only relevant to ease of drilling.
Rock failure is normally in shear; unconfined compression in laboratory test produces oblique failure shears.
Compressive strength of most rocks > applied engineering stresses; exceptions are weak clay, and any heavily weathered or densely fractured rock.
(UCS concrete = 40 N/mm^2 = 40 MPa)

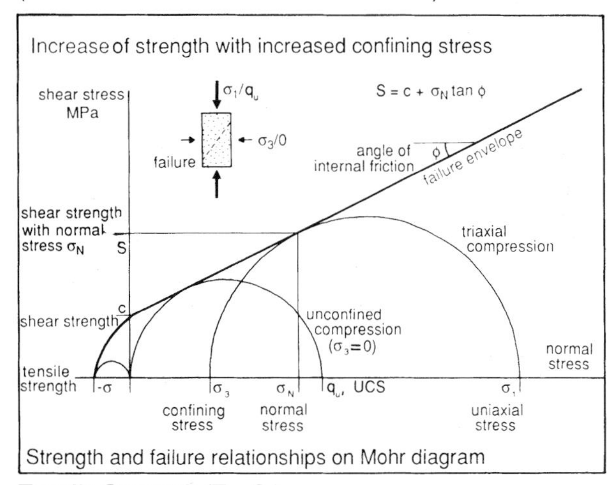

Strength and failure relationships on Mohr diagram

Tensile Strength (T_O, S_t)

Rarely measured or applied directly.
Generally about UCS/20 to UCS/8 for rocks.
Flexural strength relates to tensile strength on the outer surface, and is not easily measured or defined.
Elastic mica plates give slate high flexural strength.

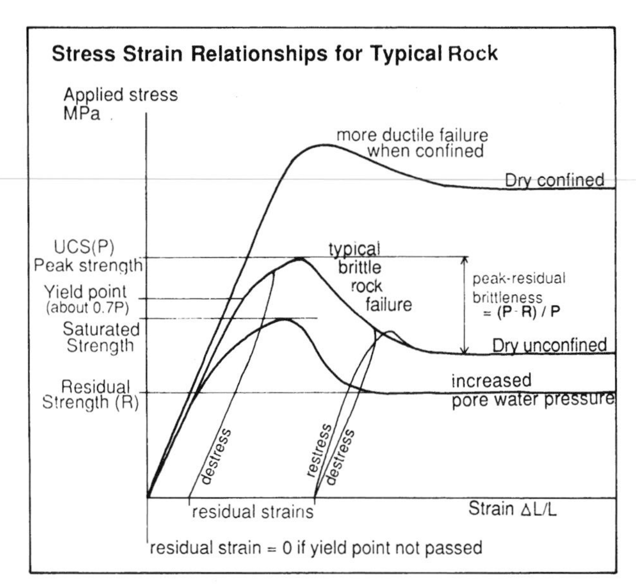

Unconfined Compressive Strength (q_u, UCS)

Strength under uniaxial load in unconfined state.
UCS of dry rock is standard for defining rock strength.
Broadly relates to porosity, and therefore to dry density.
Most igneous rocks have porosity <1%, UCS > 200 MPa.
Sedimentary rocks with density < 2·3 t/m^3 generally have UCS < 70MPa.
UCS increases with age in most sedimentary rocks due to increased lithification and reduced porosity.

Modulus of Elasticity (E)

Stress increment per strain increment, therefore directly related to strength. Known as Young's modulus.
Ductile failure starts where confining stress > UCS.
Modulus Ratio is E/UCS. Around 300 for most rocks; >500 for some strong, stiff limestones; <100 for deformable rocks, clays, some shales.

Strength Properties of Rocks

rock type	density dry t/m^3	porosity %	dry UCS range MPa	dry UCS mean MPa	UCS saturated MPa	modulus of elasticity GPa	tensile strength MPa	shear strength MPa	friction angle $\phi°$
Granite	2·7	1	50–350	200		75	15	35	55
Basalt	2·9	2	100–350	250		90	15	40	50
Greywacke	2·6	3	100–200	180	160	60	15	30	45
Sandstone – Carboniferous	2·2	12	30–100	70	50	30	5	15	45
Sandstone – Triassic	1·9	25	5–40	20	10	4	1	4	40
Limestone – Carboniferous	2·6	3	50–150	100	90	60	10	30	35
Limestone – Jurassic	2·3	15	15–70	25	15	15	2	5	35
Chalk	1.8	30	5–30	15	5	6	0·3	3	25
Mudstone – Carboniferous	2·3	10	10–50	40	20	10	1		30
Shale – Carboniferous	2·3	15	5–30	20	5	2	0·5		25
Clay – Cretaceous	1·8	30	1–4	2		0·2	0·2	0.7	20
Coal	1·4	10	2–100	30		10	2		
Gypsum	2·2	5	20–30	25		20	1		30
Salt	2·1	5	5–20	12		5			
Hornfels	2·7	1	200–350	250		80			40
Marble	2·6	1	60–200	100		60	10	32	35
Gneiss	2·7	1	50–200	150		45	10	30	30
Schist	2·7	3	20–100	60		20	2		25
Slate	2·7	1	20–250	90		30	10		25

These are mean or typical values, which can only be taken as approximate guidelines.
All values refer to intact rock which has not been weakened by weathering. Unquoted values indicate extreme variation related to orientation etc., or lack of adequate data.

Sedimentary rocks become stronger with age and tectonic stress; these values are typical for Britain and eastern USA; most rocks of similar ages are stronger in areas of plate boundary deformation such as Alpine Europe and western America.
Values of ϕ are for intact rock, ignoring fractures.

SHEAR STRENGTH OF ROCKS

May be regarded as having two components:

- cohesion (and tensile strength) due to interlocking
- internal friction, increasing under confining load

Confined Triaxial Strength

Rock strength greatly increases where confined in the ground, to values generally beyond significance to engineering loading.
Triaxial testing relates shear strength to normal stress.
Rarely measured in rocks (but important for soils).

Angle of Internal Friction ϕ

Relates confined shear strength to applied normal load, by the Coulomb equation: $s = c + \sigma_n \tan\phi$, meaning

- shear strength = cohesion + normal stress x tan ϕ

Shear Strength (S_i, S_s)

Resistance to direct shear when unconfined.
General relationship applies: $UCS = 2S_s \tan(45 + \phi/2)$.
S_s varies UCS/6 in strong rock, to UCS/2 in soft clay.
Peak strength on initial shearing declines to residual strength along the sheared surface; there is no accepted measure of rock brittleness (post-peak decline of strength).
Shear strength equates with cohesion (c) of soils.

EFFECTS OF WATER

The presence of water and any increased pore water pressure significantly reduce rock strength.
Water interrupts the bonding between minerals, and allows the break-up of clay cements in some sedimentary rocks.
Pore water pressure acts in opposition to confining stress; this reduces effective normal stress in triaxial situation, and therefore reduces confined shear strength. Important in clays and soils.
Saturation slightly reduces ϕ and greatly reduces apparent cohesion.
Water greatly reduces strength of weak, porous sedimentary rocks, but has minimal effect on strong rocks with low porosity.

Strength Recognition and Description

Rock/soil description	UCS (MPa)	Field properties
Very strong rock	>100	firm hammering to break
Strong rock	50–100	break by hammer in hand
Moderately strong rock	12·5–50	dent with hammer pick
Moderately weak rock	5·0–12·5	cannot cut by hand
Weak rock	1·5–5·0	crumbles under pick blows
Very weak rock	0·6–1·5	break by hand
Very stiff soil	0·3–0·6	indent by fingernail
Stiff soil	0·15–0·3	cannot mould in fingers
Firm soil	0·08–0·15	mould by fingers
Soft soil	0·04–0·08	mould easily in fingers
Very soft soil	<0·04	exudes between fingers

STRENGTH TESTING

Laboratory tests of rock strength suffer because of rock variation (notably in weaker sedimentary rocks), so all values recognize error of ± 20%.
Also, tests of intact rock ignore the fractures which dominate the level of rock mass strength.
In practice, it is therefore often adequate to identify the rock and read strength values from tables.

Unconfined Compressive Test

Cube or cylinder of rock with flat, cut, parallel faces, loaded uniaxially between flat steel platens; sample diameter ≥ 54 mm.
Most common and easiest test of rock strength.

Triaxial Test

Cylinder of rock loaded axially (σ_1) with equal confining stresses on radial axes due to fluid bath pressure (σ_3).
Plot on Mohr diagram to determine ϕ and c.

Ring Shear Test

Transverse shear on 2 surfaces across rock cylinder.
One of a number of shear tests, which can also be applied with confining pressure to determine ϕ.
Generally restricted to soils and weak rocks.

Brazilian Test

Cylinder of rock loaded across its diameter between two flat steel platens.
Easier than direct tensile test.

Point Load Test

Cylinder of rock loaded across its diameter between two 60° steel points with tip radius of 5 mm.
Standard portable apparatus ideal for rapid, direct field testing of borehole cores.
Can also apply multiple tests on irregular rock lumps with dimensions close to 1:1:2.
Use 54 mm core or apply correction factor (as larger core gives lower values) and ignore low results due to fracture failures.
Point Load Strength (I_s) is then close to UCS/20.

Schmidt Hammer

Hand held, spring loaded hammer measures rebound from rock surface; rebound values correlate with UCS and decline significantly in fractured rock.

Schmidt hardness	20	30	40	50	60
UCS (MPa)	12	25	50	100	200

Very rapid field test may identify weaker or weathered rock, or loose fracture blocks, in exposed rock face.

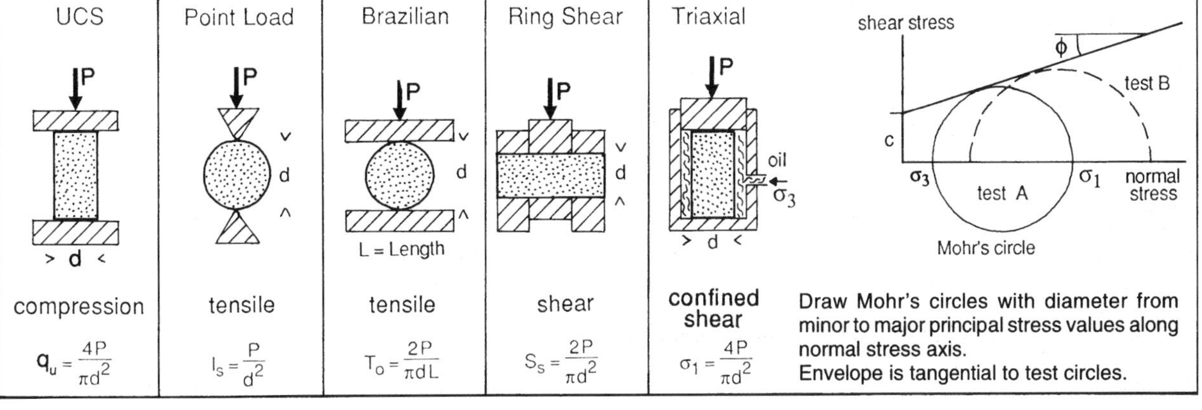

Draw Mohr's circles with diameter from minor to major principal stress values along normal stress axis.
Envelope is tangential to test circles.

25 Rock Mass Strength

Strength of a rock mass largely depends on the density, nature and extent of the fractures within it.
Rock mass strength also relates to rock strength, weathering and water conditions.

FRACTURE DENSITIES

Rock fractures include microfissures (spacing mostly 1mm–1cm), joints (1cm–1m) and faults (>1m).
Also bedding, cleavage, schistosity.
Fractures allow inelastic deformation and reduce rock mass strength to 1/5 to 1/10 of the intact rock strength. This fraction may be known as the Rock Mass Factor.
Assessing fracture density is subjective, except by RQD.

Rock Quality Designation (RQD) is a fracture quantification on borehole core > 50 mm diameter; lengths of core pieces are measured as they come from the drill barrel, and:

- RQD = Σ(core lengths >10 cm) × 100/borehole length

Values of RQD > 70 generally indicate sound rock.

FRACTURE ORIENTATION

Influence of orientation is only assessed subjectively in terms of favourability with respect to potential failure by sliding or rotation at a particular site or part of a site.
Importance of orientation is shown by UCS variation in blocks of slate with well-defined cleavage.

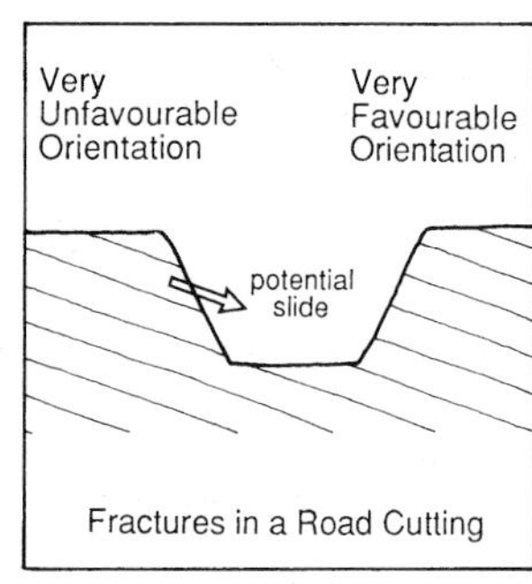

Fractures in a Road Cutting

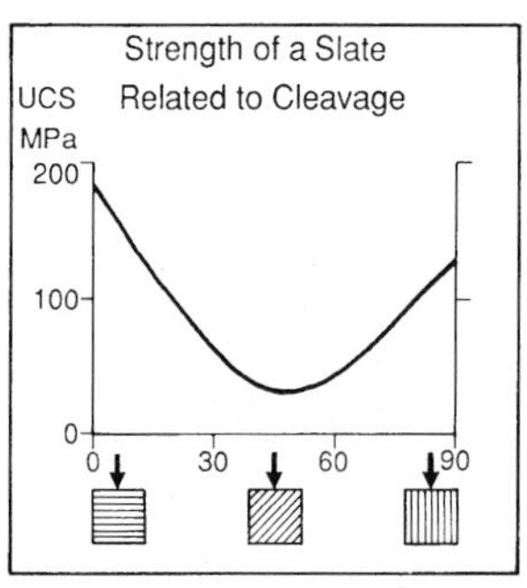

Strength of a Slate Related to Cleavage

FRACTURE TYPES

Fracture roughness influences its shear strength.
Shear of very rough fractures requires dilation of rock mass as irregularities override each other.
Roughness is difficult to assess and quantify.
Fracture infills include clay gouge, clay from weathering, breccias, and slickensided gouge.

Typical friction angles (ϕ): clean rock 20–50°; clay fill 10–20°; breccia 25–40°

Cohesion across fractures varies 0–500 kPa.

ROCK MASS CLASSIFICATION

Assessment of rock mass strength recognizes cumulative effect of different geological features.
Classification is therefore an accumulation of weighted values given to selected parameters.
Two most widely used systems are Geomechanics RMR system which adds rating values (below), and Norwegian Q system which multiplies rating values. Both systems are dominated by fracture properties.
Applications to specific engineering problems where rock mass class gives approximate guideline values of ground parameters, as in the lower table.

Geomechanics System of Rock Mass Rating (RMR) uses parameters and point scoring in table below.

Norwegian Q System successfully multiplies rating values to determine the rock mass quality (Q) as:

$$Q = (RQD/Jn) \times (Jr/Ja) \times (Jw/SRF)$$

Factors with rating ranges from good to bad are:

RQD	= Rock quality designation	100–10
Jn	= Joint set number	1–20
Jr	= Joint roughness factor	4–1
Ja	= Joint alteration and clay fillings	1–20
Jw	= Joint water inflow or pressure	1–0·1
SRF	= Stress reduction factor due to excavation	1–20

Q values range from < 0·01 to > 100.
System is tabulated in the appendix (page 82)

Geomechanics System of Rock Mass Rating

Parameter	Assessment of values and rating				
Intact rock UCS, MPa	>250	100–250	50–100	25–50	1–25
Rating	15	12	7	4	1
RQD %	>90	75–90	50–75	25–50	< 25
Rating	20	17	13	8	3
Mean fracture spacing	> 2 m	0·6–2 m	200–600 mm	60–200 mm	< 60 mm
Rating	20	15	10	8	5
Fracture conditions	rough tight	open < 1 mm	weathered	gouge < 5 mm	gouge > 5 mm
Rating	30	25	20	10	0
Groundwater state	dry	damp	wet	dripping	flowing
Rating	15	10	7	4	0
Fracture orientation	v. favourable	favourable	fair	unfavourable	v. unfavourable
Rating	0	–2	–7	–15	–25
Rock mass rating (RMR) is sum of the six ratings			Note that orientation ratings are negative		

Guideline Properties of Rock Mass Classes

Class	I	II	III	IV	V
Description	very good rock	good rock	fair rock	poor rock	very poor rock
RMR	80–100	60–80	40–60	20–40	< 20
Q Value	> 40	10–40	4–10	1–4	<1
Friction angle ϕ (°)	> 45	35–45	25–35	15–25	<15
Cohesion (kPa)	> 400	300–400	200–300	100–200	< 100
SBP (MPa)	10	4–6	1–2	0·5	< 0·2
Safe cut slope (°)	> 70	65	55	45	< 40
Tunnel support	none	spot bolts	pattern bolts	bolts + shotcrete	steel ribs
Stand up time for span	20 yr for 15 m	1 yr for 10 m	1 wk for 5 m	12 h for 2 m	30 min for 1 m

Foundations on Rock

SAFE BEARING PRESSURES

Guideline values for maximum loads that may safely be imposed on undisturbed ground.

May be estimated in many ways, all based on past experience and incorporating ample safety factors to allow for variable ground conditions.

Values are useful preliminary design guides, as it is normally uneconomic to complete meaningful field tests on fractured rock masses.

May be based on rock type:

Safe Bearing Pressure – typical values

Rock types	Unweathered and massive	Heavily fractured or thinly bedded
Strong igneous rock, gneisses	10 MPa	6 MPa
Strong limestones and sandstones	4 MPa	3 MPa
Schists and slates	3 MPa	2 MPa
Strong mudstones, soft sandstones	2 MPa	1 MPa
Shale, sound chalk, soft mudstone	750 kPa	400 kPa

Or based on rock strength and fracturing:

Safe Bearing Pressure – guidance values

SBP (MPa):

UCS (MPa)	RQD 25 / Fracture spacing 60	RQD 70 / Fracture spacing 200	RQD 90 / Fracture spacing 600 mm
100	4	8	12
25	1	3	5
10	0·2	1	2

Improved estimates of SBP can take account of the rock modulus ratio (E/UCS); less deformable rocks with high modulus ratios, such as limestone or granite, can be assessed higher SBP than softer rocks, such as shale, for the same values of UCS and RQD.

SBP values are also adjusted by conventional factors with respect to foundation shape; reductions for large, shallow or cyclicly loaded foundations, are generally of little direct significance due to the high bearing capacities of rock (as opposed to soil).

Settlements on rock are generally small, and are rarely constraining influences once bearing capacity criteria have been satisfied. (They are normally the limiting factor for foundations on clay soils.)

UPB = Ultimate Bearing Pressure = load at failure

SBP = Safe Bearing Pressure = UBP ÷ Safety Factor, usually 3; similar to Presumed Bearing Value quoted for soils.

ABP = Acceptable Bearing Pressure = SBP further reduced to satisfy specific structural requirements such as settlement; reduction factor may be significant on soils, usually close to 1 for rocks.

FOUNDATIONS FOR TORONTO CN TOWER

World's tallest free-standing structure is 550 m high, weighing 110 000 t.

Founded on shale, UCS = 10–25 MPa, E = 3·7 GPa, RQD = 50–80, with some thin weak bands which were mapped and avoided.

Slab foundation 7 m below rockhead, beneath 10 m drift.

Mean load on shale is 580 kPa, with peak stress in high wind of 2·89 MPa; compared to design SBP for deep caissons of 7·2 MPa in the same rock.

Settlement was 6 mm, after 6 mm heave in excavations.

FAILURE OF ROCK

Sound rock is capable of bearing most normal engineering loads; the same cannot be said for soils.

Normal variations in rock properties are covered by generous factors of safety in engineering design.

Major zones of significant weakness, including underground voids, can cause failures; should be avoided by adequate site investigation.

There are four possible modes of failures:

- Shear failure and upward displacement of the rock, due to imposed loading > rock strength.
- Compaction of porous rocks (causing extreme settlement), also due to loading > rock strength.
- Rock failure into underground cavity, where rock roof fails in shear or flexural tension.
- Landsliding and lateral displacement, where slope profiles are too steep.

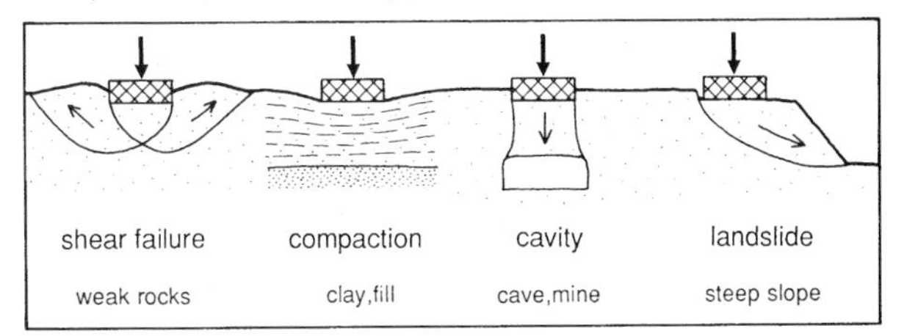

INFLUENCE OF BURIED VOIDS

Plug or beam failures of rock under structural loading over underground cavities depend on the rock strength and fracturing, the cavity size and depth, and the applied loads and stresses.

Natural and mined cavities can vary greatly in size, shape and stability, and each one requires individual assessment if it is relevant to engineering works.

Risk of ground failure increases if any one of the following guideline criteria is met:

- Cover thickness < cavity width;
- Cover thickness underneath end bearing piles < 5 times pile diameter;
- Loading to SBP above < 3 m of strong rock;
- Cover of weak rock or soil (with progressive failure and cavity migration) < 10 times cavity height.

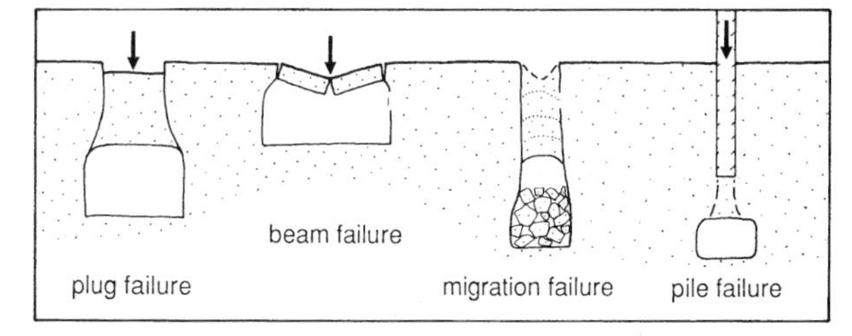

GROUND IMPROVEMENT

Treatment of fresh rock is rarely necessary or economic for structural foundations.

Weathered and weak rock near surface is better removed or piled through.

Injection of cement grout to fill rock pores and increase strength is limited by low permeability of intact rock.

Grouting can double mass strength of fissured rocks. Underground cavities can be filled with grout injected through 100 mm boreholes; may need 3–4 m grid of holes if cavities are partially blocked; use fluid mixture of 1:10 ratio of cement:PFA or fines; need stiff grout with sand or gravel to form perimeter barrier to avoid high losses away from site.

Alternatives to cement grout are foamed concrete or uncemented rock paste if need is only to prevent progressive roof falls.

26 Soil Strength

Properties of a soil depend on the grain size, mineralogy and water content, all of which are inter-related.
Clay minerals can hold high water content; for fine grained soils, critical concept is consistency related to water content.

SOIL CONSISTENCY

With varying water content, a soil may be solid, plastic or liquid. Most natural clays are plastic.
Water content (w) = weight of water as % of dry weight.
Consistency limits (Atterberg limits) are defined as:
Plastic limit (PL) = minimum moisture content where a soil can be rolled into a cylinder 3 mm in diameter.
Disturbed soil at PL has shear strength around 100 kPa.
Liquid limit (LL) = minimum moisture content at which soil flows under its own weight.
Disturbed soil at LL has shear strength around 1 kPa.
Plasticity index (PI) = LL–PL. This refers to the soil itself and is the change in water content required to increase strength 100 times; it is the range of water content when the soil is plastic or sticky.
High PI soils are less stable, with large swelling potential.
Liquidity Index (LI) = (w-PL)/PI. This is a measure of soil consistency and strength at a given water content.

Clay Mineral	Activity	PI	ϕ
Kaolinite	0·4	30	15
Illite	0·9	70	10
Smectite	>2	400	5
PI values are for soil with 75% clay fraction			

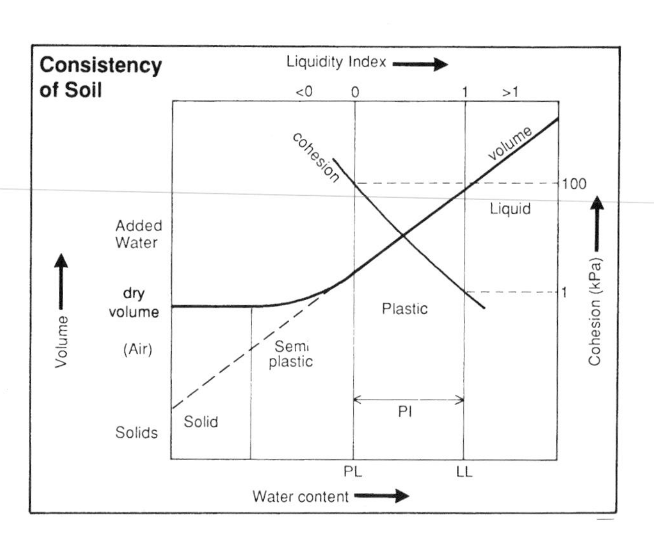

CLAY MINERALS

Plasticity and properties of clay soils depend on amount and type of clay minerals.
Soils with < 25% clay minerals are generally stronger, with low PI and ϕ < 20%.
Activity of clay = PI / % fines (< 0·002 mm diameter).
Soils with high clay fraction and high activity can retain high water content, giving them low strength, and also have low permeability.
Activity is mainly due to clay mineral type; smectite (montmorillonite) clays are the most unstable.

SOIL CLASSIFICATION

Soils are classified on grain size and consistency limits.
A-line distinguishes visually similar clays and soils.
More subdivisions exist in a full soil classification.

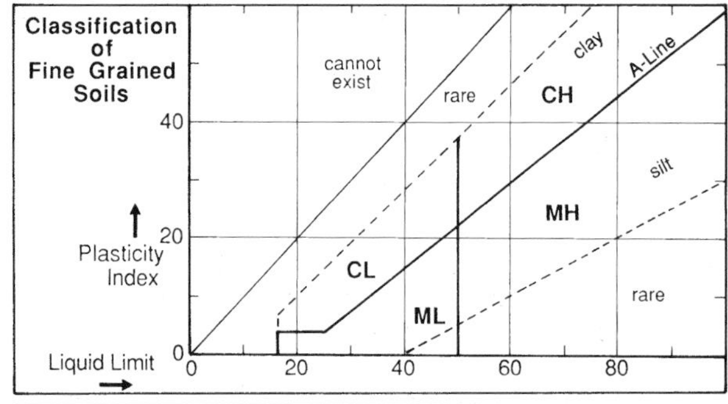

SHEAR STRENGTH

All soils fail in shear.
Shear strength is a combination of cohesion and internal friction; expressed by Coulomb failure envelope.
Cohesion (c) derives from interparticle bonds; significant in clays, zero in pure sands.
Angle of internal friction (ϕ) is due to structural roughness; higher in sand than in clay.

- Shear strength = cohesion + normal stress x tan ϕ

Normal stress is critical to shear strength but pore water pressure (pwp) carries part of overburden load on soil, thereby reducing normal stress.

- Effective stress (σ') = normal stress (σ) – pwp.

Shear strength is correctly defined in terms of effective stress, so that:

- Shear strength (τ) = c' + σ'tan ϕ'

Soil Classification		grainsize	typical values		
type	class	mm	LL	PI	ϕ
Gravel	G	2–60			>32
Sand	S	0·06–2			>32
Silt	ML	0·002–0·006	30	5	32
Clayey silt	MH	0·002–0·06	70	30	25
Clay	CL	<0·002	35	20	28
Plastic clay	CH	<0·002	70	45	19
Organic	O	–			<10

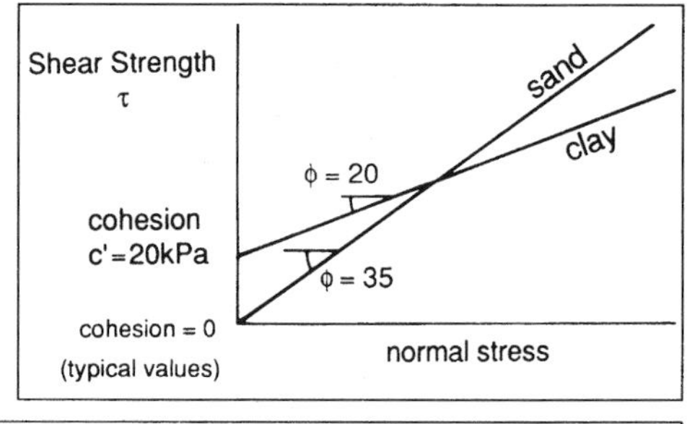

Properties of Cohesive Clay Soils							
Material	State	LI	SPT, N	CPT, MPa	c, kPa	m_v, m²/MN	ABP, kPa
Alluvial clays	soft	>0·5	2–4	0·3–0·5	20–40	>1·0	<75
	firm	0·2→0·5	4–8	0·5–1	40–75	0·3–1·0	75–150
Till and Tertiary clays	stiff	–0·1→0·2	8–15	1–2	75–150	0·1–0·3	150–300
	v. stiff	–0·4→–0·1	15–30	2–4	150–300	0·05–0·1	300–600
	hard	<–0·4	>30	>4	>300	<0·005	>600

Cohesion (c) is equivalent to short term shear strength

STRENGTH DECLINE IN CLAYS

Drainage progress of a loaded clay is critical as any increase of pore water pressure may lead to failure; significant in new excavations and embankments.

Peak strength declines to residual strength due to restructuring, notably alignment of mineral plates, during dislocation along a plane. Change is due to almost total loss of cohesion and also reduction in friction angle. Significant in all clays, notably those with higher PI.

- Brittleness = % decline from peak strength.

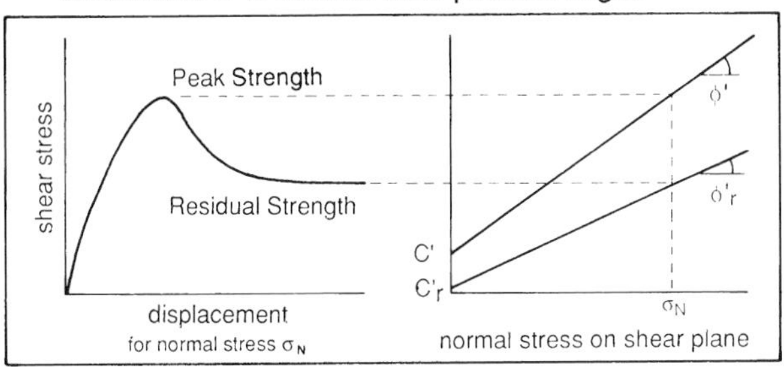

Sensitive clays lose great proportion of their strength on restructuring of entire mass; they have high LI and small grain size, so cannot drain rapidly and load is taken by pwp; shear strength approaches zero.

- Sensitivity = ratio of undisturbed:disturbed strengths and relates to undrained brittleness.

CONSOLIDATION

This is decrease of volume, under stress.

Primary consolidation is large and fast; due to expulsion of water until excess pwp is zero.

Secondary consolidation is small and slow; due to restructuring and lateral movement; same as drained creep.

Normally consolidated clays are those compacted by their present overburden of sediments.

Over-consolidated clays are those more compacted in the past by overburden soils since removed by erosion (or by glacier ice); they can bear loading up to their previous overburden stress with only minimal compression and settlement.

Compression coefficient = m_v = reduction of thickness with increase of stress; correlates closely with LL.

CONE PENETRATION TEST (CPT)

In a site investigation borehole, a 60° cone (= 36 mm in diameter) is driven into soil at 15–25 mm/second, followed by a concentric outer sheath.

End resistance and sheath resistance are measured: Friction ratio = (side friction/end friction) | 100;

ratios on standard electrical systems differ on less commonly used mechanical systems.

Values relate to soil types and packing state, and give indication of Acceptable Bearing Pressure.

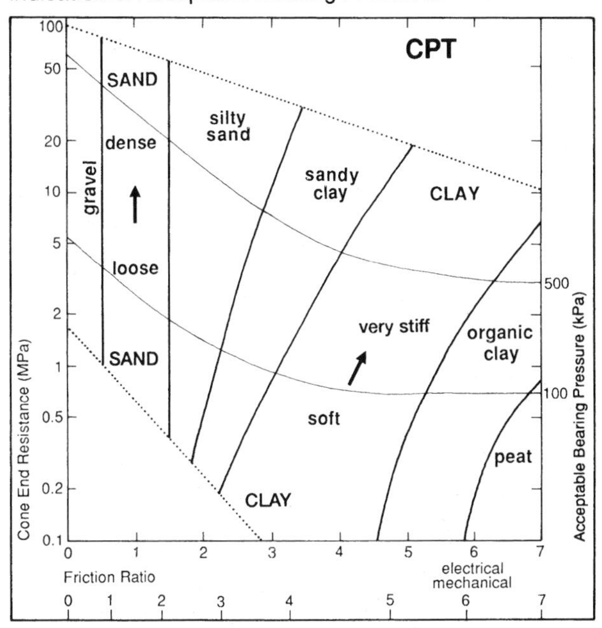

ACCEPTABLE BEARING PRESSURE

Values relate largely to soil water content and consolidation history.

Depend on SBP and acceptable settlement.

- Settlement = $m_v \times$ thickness $\times$ imposed stress.

Rate of settlement depends on permeability; slow in clay soils which cannot drain rapidly.

Settlements on clay may be large: then referred to as subsidence, along with other processes which affect clays (section 28).

Non-cohesive Soils

Sand soils, and gravels have no cohesion, except that derived from any clay matrix and water suction. Sand stands in steep slopes when wet due to negative pore pressure (critical in building sand castles), but will not stand when dry or saturated.

Strength, slope stability and bearing capacity all derive from internal friction; ϕ for granular soils (sands and gravels) range 30–45°; increases due to grading, packing density and grain angularity.

Settlement is small and rapid; not usually considered, except on very loose sands and artificial fills.

Properties are best assessed in situ by SPT; N values are a function of packing density and grading.

Bearing capacity of sandy soils may be improved by dynamic consolidation (with a 20 ton weight repeatedly dropped from a crane) or by vibrocompaction.

Properties of Sands

Packing	RD	SPT	CPT	ϕ	SBP
v. loose	<0·2	< 5	<2	<30	<30
loose	0·2–0·4	5–10	2–4	30–32	30–80
m. dense	0·4–0·6	11–30	4–12	32–36	80–300
dense	0·6–0·8	31–50	12–20	36–40	300–500
v. dense	>0·8	>50	≥20	≥40	≥500

STANDARD PENETRATION TEST (SPT)

In a site investigation borehole, a 51 mm split tube sampler is driven for 150 mm.

Using 64 kg hammer dropped 760 mm, number of blows (N) is counted to drive the tube the next 300 mm.

A simple, effective test; N values closely relate to sand properties; should be used with care in clay soils.

(At shallow depth N may be multiplied by empirical correction factor, F, to allow for low stress;

$F = 350/(25D + 70)$, where D = depth in m.)

Relative Density is a measure of grain packing on a scale from loosest to densest possible states of compaction.

SPT refers to corrected N values.

CPT values are end resistances, in MPa, for fine sand; values are lower in silts and lighter in gravels.

Friction angles are for average sand; add 2° for angular grains; subtract 3° for rounded grains; add 5° for gravels.

SBP values, in kPa, are for foundations 3 m wide with settlement < 25 mm; multiply by 1·4 for strip foundations 1 m wide; values are halved for sand stressed below water table.

27 Ground Subsidence

Subsidence is only possible where the ground material can be displaced into some sort of underground voids, which can only occur in certain rock types.
Macrovoids, large cavities: solution caves in limestones (section 29); much rarer natural cavities in other rocks, including salt and basalt; mined cavities in any rocks of economic value (sections 30, 31).
Microvoids in very porous, deformable rocks: most important in clay (section 28); in peat, some silts and some sands; in made ground and backfill (section 30).

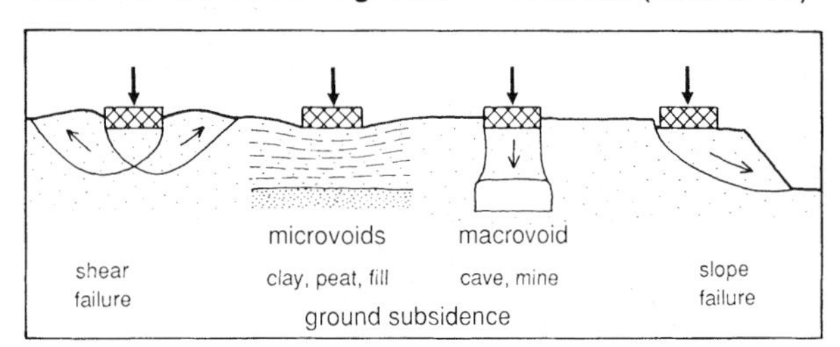

Subsidence cannot occur on solid, unmined rock – sandstone, granite, mudstone, slate – except by shear failure and rotational displacement to the surface under excessive load, or by landsliding where slope profiles permit (section 32).
Hazard of potential subsidence can therefore be recognized by rock type on geological maps.
All rocks do compact under load. Weak mudstone or sandstone can compact enough to cause settlement of structures, but normally well inside acceptable limits.

COLLAPSIBLE SOIL HYDROCOMPACTION

Some fine soils collapse due to restructuring when saturated for the first time; this hydrocompaction may cause subsidence by 15% of the soil thickness. The collapse is due to total loss of cohesion, after disruption of fragile clay bonds or solution of a soluble cement.
Loess collapses most easily where it contains about 20% clay; with more or less clay, it is less unstable.
Alluvial silts deposited by flood events in semi-arid basins, some tropical soils and some artificial fills may all exhibit collapse on saturation.
Collapse potential is highest in soils with dry density $< 1{\cdot}5$ t/m^3, liquid limit < 30, and moisture content $< 15\%$ in dry climate zones. Potential can be recognized by consolidation test with saturation part of the way through the loading cycle.

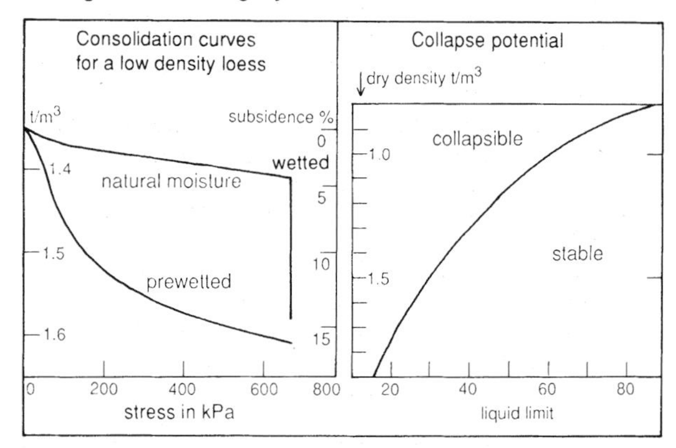

Some collapsing soils only hydrocompact with loading.
Subsidence hazard is highest in irrigated arid areas, e.g. Central Valley of California.
Soil collapse may be induced prior to construction by pre-wetting through flooding; thin soils respond to dynamic consolidation or vibroflotation.

SALT SUBSIDENCE

Rock salt may occur as extensive beds in sedimentary sequences. It dissolves in circulating groundwater rapidly enough to cause slow natural subsidence.
Most solution takes place at the rockhead beneath permeable drift; thus creates a residual breccia of the collapsed mudstone which was interbedded with the salt; cavities collapse before they become large.
Linear subsidences are localized over 'brine streams' – zones of concentrated groundwater flow along rockhead, commonly guided along salt band outcrops; typical subsidences are 5 m deep, 100 m wide, 5 km long.

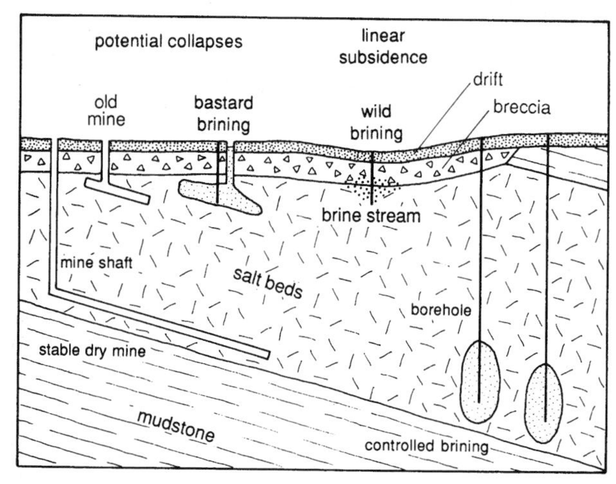

Wild brining is uncontrolled pumping from the brine streams; it greatly accelerates formation of the linear subsidences, which may form in tens of years.
Deep solution mining (controlled brining) and modern deep mines in dry salt are both stable: no subsidence.
Pumping brine from old shallow mines (bastard brining) causes serious collapses; now illegal in Britain.
Most surface movements are small and slow; engineering precautions are similar to those for longwall mining (section 31).
Cheshire has the worst salt subsidence in Britain; houses and structures in Northwich all have timber or steel frames or concrete rafts that can be jacked up. Now that wild brining has almost ended, subsidence due to natural solution is very slow – but does continue.

GYPSUM SOLUTION

Gypsum may be dissolved and removed naturally.
Solution is slower than of salt, faster than of limestone – rock can dissolve within the lifetime of a built structure.
Rockhead pinnacles may be dissolved by groundwater, so may not be safe for foundations in the long term.
Caves are smaller and less common than in the strong limestones, but may create a significant hazard where weak roof rock collapses easily to create sinkholes.
Plugging or filling cavities in gypsum requires care, as diverted groundwater may rapidly create new caves.

NATURAL CAVES

Common in limestone and gypsum; rare in other rocks.
Basalt may contain lava tubes on shield volcanoes.
Open fissures hidden beneath soil cover may develop by slope movement – round heads of landslides and as gulls on camber folds.
Soil pipes, sea caves and rock arches are all of limited extent; latter are conspicuous as surface features.

SUBSIDENCE ON PEAT

Peat may contain ten times its own weight of water; it can shrink by 10–75% under load.
When loaded to exceed its very low shear strength, peat also creeps and spreads; so very high settlements are normal; coefficient of compressibility, $m_v > 1.5\ m^2/MN$.

Drainage of peat causes surface subsidence of up to 60% of the groundwater head decline; less on later redrainage.

Wastage, by oxidation of biomass above watertable, continues at low rate dependant on climate; causes surface lowering, and major loss of agricultural land; reduced by maintaining high water table.

Strength of undrained peat is negligible, drained peat may be UCS = 20–30 kPa, and E = 100–140 kPa.

Peat consolidated by structural load gains strength; may reach SBP = 50–70 kPa. Primary consolidation takes place in days; secondary stage may last years. Laboratory testing and consolidation prediction are hindered by variability of peat and difficulties of sampling; full scale field tests may be worthwhile for major projects.

CONSTRUCTION ON PEAT

Removal is economical if peat is less than about 3 m thick. Displacement of thicker peat is possible by end-tipped sand, purely by gravity, or aided by jetting to 6 m deep, or peat-blasting to 9 m deep.
Piles through peat are often economic, and required by state law in some of USA; house foundations may be left above ground if drained wastage continues.
Pre-loading is successful with surcharge of 1–3 m of sand or fill for 1–12 months; rebound is about 5%.
Sand drains are of limited use as peat permeability is high; wick drains have been used to accelerate consolidation in English Fenlands.
Embankments on peat may cause more settlement than their height. So lightweight fill is used; polystyrene blocks are best; sawdust, brushwood and peat bales have been used in Canada and Ireland, and are stable when depressed below water table.
Rafts can be used for light, centrally loaded buildings, with underrim to reduce peat spreading; houses on rafts in northern England settled 800 mm on 2·5 m of peat with imposed load of only 15 kPa.
Basements to give nil net loading rarely economic for houses.

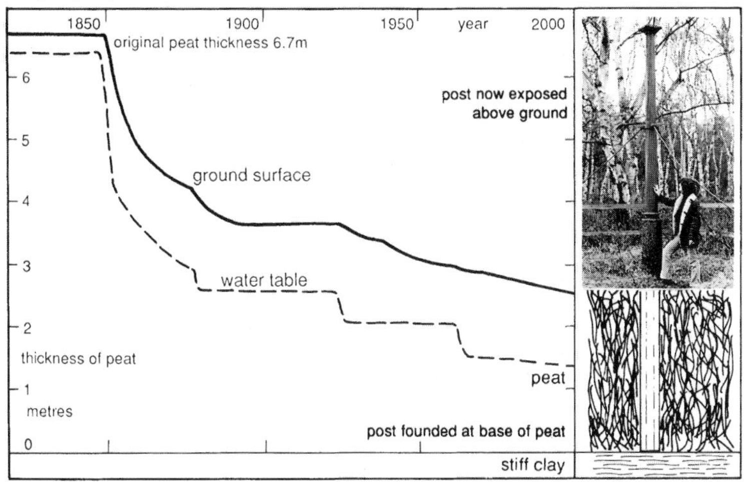

Pumped drainage and ground subsidence recorded over 150 years against the Holme Post in the peat of the English Fenlands

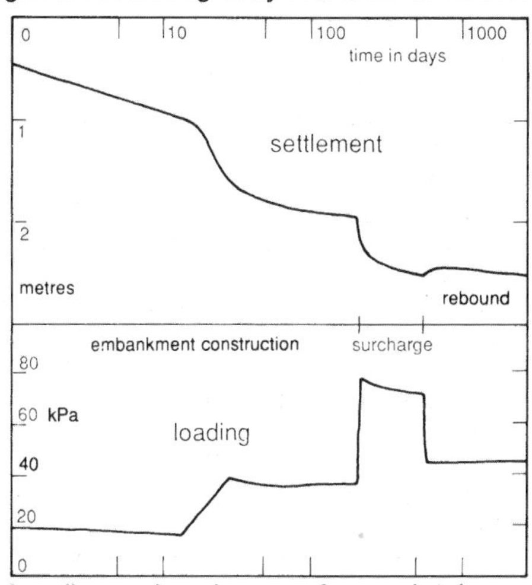

Loading and settlement of an embankment for a road over peat in Canada

EARTHQUAKE LIQUEFACTION

Sand may liquefy due to a temporary loss of effective stress during period of earthquake vibration, if it is:

- Uniformly graded, with grain size < 0·7 mm
- Poorly packed with low relative density
- Below the water table at shallow depth.

Hazard zones may be defined by SPT, notably where N-values < 20 at 10 m depth.
Liquefaction causes total loss of strength during the period of vibration, as in the 1964 earthquake at Niigata, Japan, when buildings subsided rapidly into saturated alluvial sands.
Stabilize sand and reduce hazard by dynamic consolidation, drainage and water table decline, or surcharge to raise internal stress.

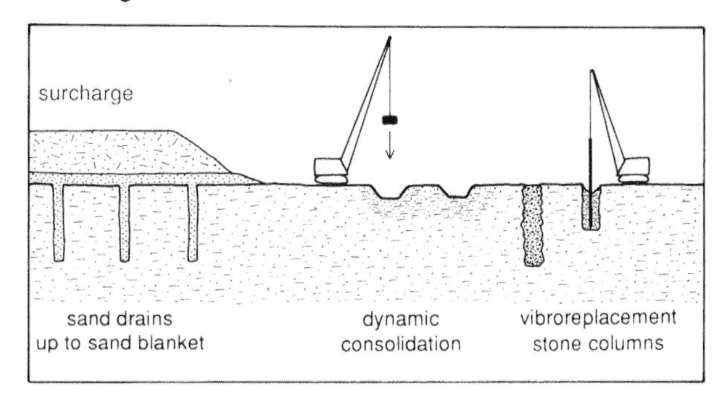

GROUND IMPROVEMENT

Surcharge: consolidation accelerates under a few metres of placed fill and almost stops when surcharge is removed, usually after one year, pre-construction.
Drainage: accelerates water expulsion, so accelerates consolidation; may allow settlement beneath embankment to be completed during construction time. Sand or fibre drains spaced at 1–3 m most effective at depths < 15 m.
Grouting: cannot penetrate clays; 10% cement mixed into clays of LL < 45 increases strength.
Liming: adding 5% lime creates stronger soil; reduces plasticity and shrinkage; stabilizes montmorillonite by replacing sodium with calcium.
Vibrocompaction: densify sandy, non-cohesive soils with a crane-supported vibrating poker.
Vibroreplacement; feed crushed stone beside poker to create stable stone columns in cohesive soil or fill.
Dynamic consolidation: drop 15 t weight, 3–5 times, 20 m from crane, on 5–10 m grid, to densify sandy soil. May fissure a clay to aid drainage consolidation.
Ground freezing: expensive temporary stabilization of excavation.
Geotextiles: along with coarser geogrids increase shear strength, but can only be installed in placed soils, not undisturbed ground.

28 Subsidence on Clays

Clays have high porosity with deformable grains of clay mineral; so high potential compaction.
Compaction = volume decrease = consolidation. Due to water expulsion (primary consolidation) followed by restructuring (secondary consolidation).
Consolidation of clay, subsidence of surface and settlement of structures increase with imposed load or drained water loss.
Subsidence is greatest on thick clay, with high smectite content, low silt content, and of young age with minimal history of over-consolidation.
Bearing capacity of clays ranges 50–750 kPa, largely related to water content; generally limited by settlements which exceed acceptability long before threat of failure.
Older clay, shales and mudstones are stronger and less compressible; strong mudstone may have SBP = 2000 kPa; hard shales deteriorate by slaking.

SETTLEMENT

Clay is consolidated by imposed structural load.
All clays cause some degree of settlement.
Water is squeezed out by applied stress.
Subsidence of ground and settlement of structure depend on initial water content of clay and stress applied; laboratory assessment by consolidation test.
Remedy is to avoid loading the clay or to wait for settlement to stop (or reduce to acceptable rate).
Modest settlement beneath buildings may fracture brittle drains; subsequent leakage may remove mineral soil in piping failure; this also causes subsidence but involves a different process.

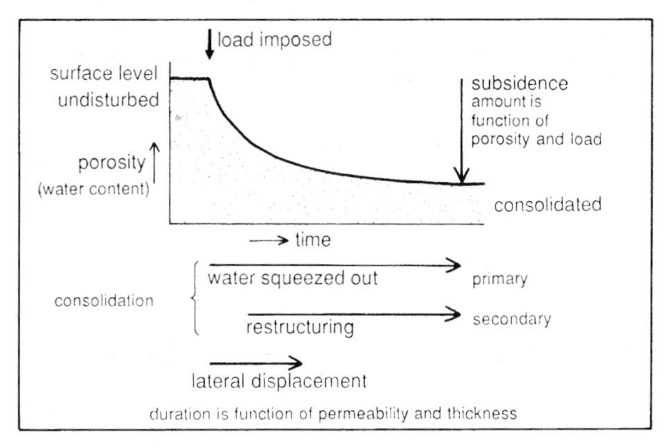

SHRINKAGE

Consolidation of clay is accelerated by water loss.
All clays exhibit some degree of shrinkage.
Water is drained out, causing volume decrease of drained soil; also loss of pore water pressure support.
Tree roots cause shrinkage in top 2 m of clay soil, but reached 6 m in London Clay in recent dry summers.
Britain's insurance claims for damage to houses on shrinkable clays are approaching £500M/year.
Pumped drainage of site may cause shrinkage nearby.
Remedy is control and stabilization of pore water pressure in clay.

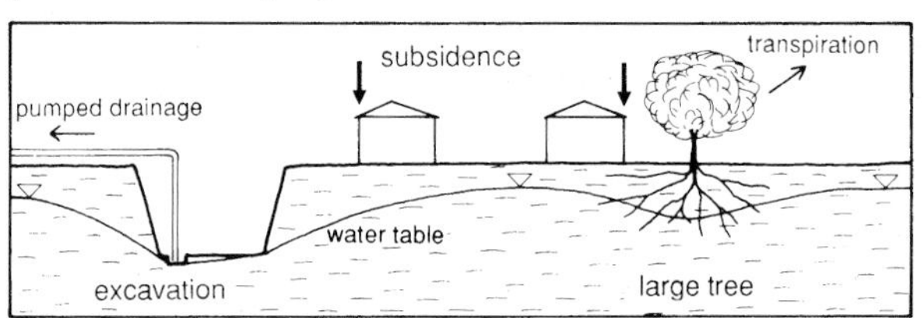

DIFFERENTIAL SETTLEMENT

Settlement of a structure most serious when differential.
Commonly due to uneven loading, lateral change of silt content in soil, rockhead slope or uncontrolled drainage.
Tilting of a tall structure creates differential loading, and then accelerates differential settlement.
Transcona grain elevator, Canada, tilted 27° in a day in 1912; clays under raft base compacted unevenly over sloping rockhead, then sheared and displaced laterally.

LEANING TOWER OF PISA

Cathedral bell tower, 58 m high, 4 m out of vertical weighs 14 000 t; imposed 500 kPa on clay with ABP ~50 kPa.
Main settlement is due to compaction and deformation of soft clay at depth of 11–22 m. Differential movement probably started due to clay variation within overlying silt layer; subsequently it was due to eccentric loading.
Stabilization in 1993–2001 was by controlled induced subsidence of north side. Temporary counter-weight, of 600 t of lead, tilted tower back 15 mm. Creep closure of 41 uncased boreholes, each 225 mm in diameter, with repeated drilling to remove a total of 35 m^3 of soil, tilted the tower back another 425 mm; so it is now stable.
Cable bracing was just for security during drilling.
Temporary tendons confined masonry to reduce risk of bursting failure until load was reduced by tilt reduction.

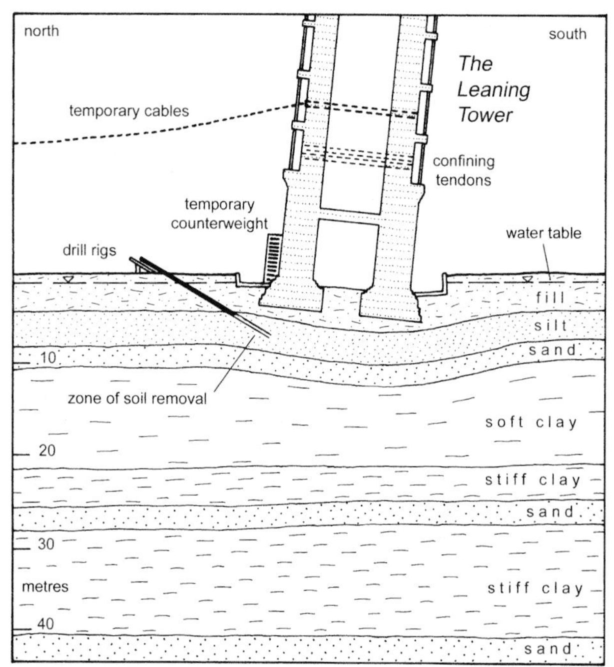

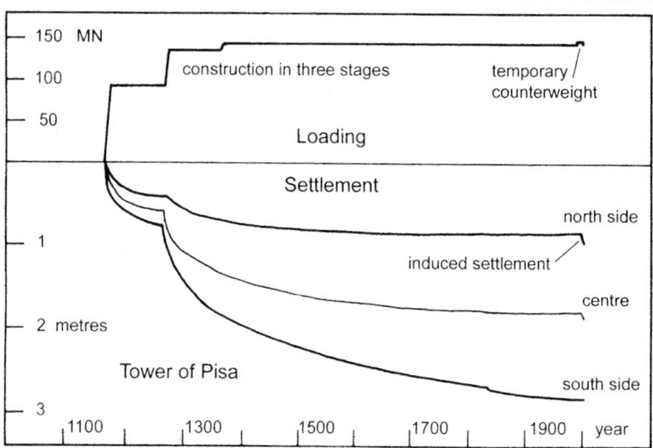

SEVERE SETTLEMENT: MEXICO CITY

City is built on drained lake bed in basin ringed by mountains of volcanic rock.
Young, porous, highly compressible clays are largely montmorillonite; water content around 300%.
All buildings on shallow foundations settle severely.
Palace of Fine Arts was built on a massive concrete raft; imposed load of 110 kPa caused 3 m settlement.
Heavy rafts create their own subsidence bowls and damage adjacent buildings.
Stable foundations are piled to sand.
Latino Americana Tower has buoyant foundations with basements to reduce imposed load, and piles to upper sand. Designed so that settlement by compaction of lower clay equals ground subsidence due to pumped head decline in upper clay (see below).

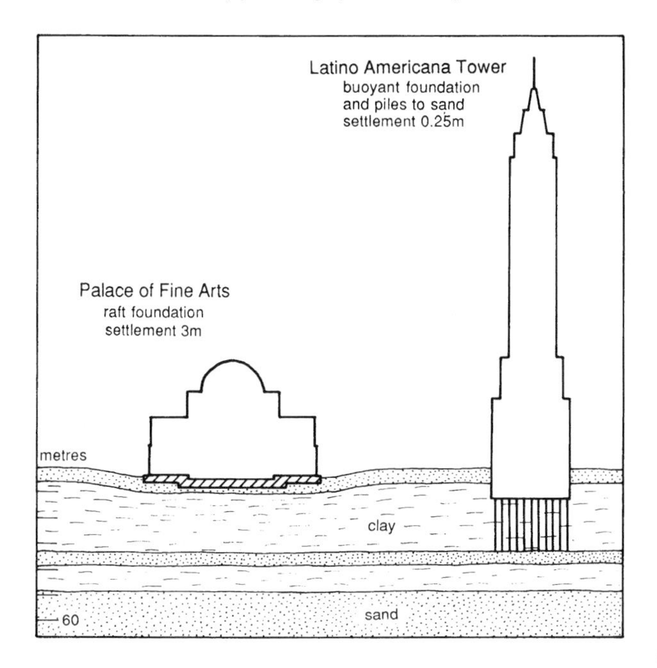

REGIONAL SUBSIDENCE

Groundwater abstraction which exceeds natural recharge causes decline of water table.
Loss of pore water pressure within clays causes widespread subsidence; significant where overpumping is from sand aquifers interbedded with clay aquitards.
Pumping from sand causes small, instantaneous, elastic, recoverable compaction of the sand.
Repressuring of aquifers has caused elastic rebound of sand – but < 10% of original subsidence.
Compaction of clay is greater, inelastic, non-recoverable; occurs as groundwater pressures equalize between sand and clay, with time delay due to low permeability of clay.
Ratio of subsidence to head loss varies with clay type:

- 1:6 on young Mexico City montmorillonite,
- 1:250 on old consolidated London Clay illite.

Subsidence stops if water tables recover.
Venice has subsided on clay; it now floods on 100 high tides per year. Subsidence has stopped since pumping of groundwater was controlled, but rising sea levels demand new barriers and raised perimeter frontage.
Mexico City has 9 m of subsidence on montmorillonite clays interbedded with over-pumped sands; founded in the sands, well casings now protrude in the streets.
Bangkok is now fastest subsiding city, at >10 cm/year.
Santa Clara Valley, California, shows correlation of water table decline with 4 m of ground subsidence, now stopped as pumping has been reduced.

EXPANSIVE SOILS

Clay soils which exhibit major free swelling on hydration and similar contraction on desiccation.
Montmorillonite is the cause – unstable clay mineral which associates with water causing crystal expansion with force of 600 kPa, but easily loses water by drainage or dessication. Sodium variety is most unstable, with liquid limit up to 500, and activity > 5; calcium variety is more stable.

> Smectite = unstable clay mineral group
> Montmorillonite = main member of smectite group
> Bentonite = clay soil with high smectite content

Montmorillonite clays form primarily by weathering of volcanic rocks in warm climates; so few expansive soils in Britain; annual costs of uplift damage on expansive soils in USA exceed combined costs of earthquakes and flooding.
Highest swelling is in any soils which are: rich in montmorillonite, fine grained, dense and consolidated, dry, remoulded, lightly loaded, with high plasticity index.
Field recognition of expansive soils: sticky when wet; polished glaze on cut dry surfaces; dry lump dropped in water expands so fast that it breaks up explosively.
Remedies for expansive soils: liming to form stable calcium variety; control of groundwater, as soils are stable if they remain wet, or are kept dry under buildings by control of drainage.

SOIL PIPING

A throughflow of water washes out the finest soil particles, so increasing the porosity, and then washes out progressively larger particles to create a pipe.
Cavity may reach a metre diameter before collapse.
Can develop naturally through terraces in silty soils.
Piping is common in any type of soil which is carried by seepage water into a broken drain.

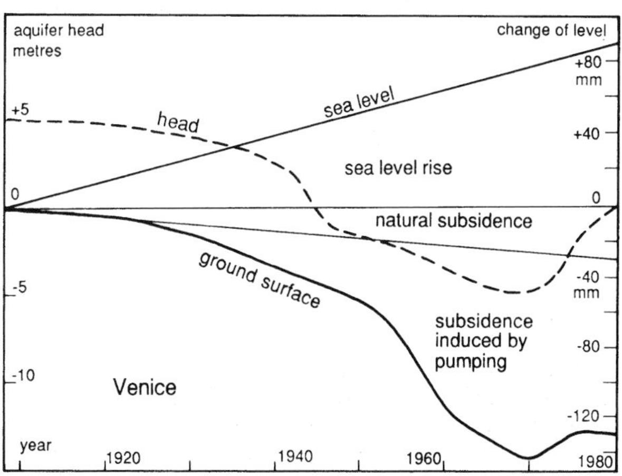

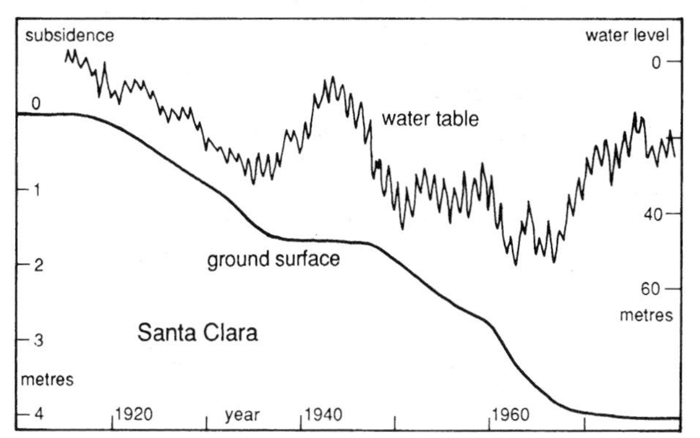

29 Subsidence on Limestone

Limestone is the only common rock soluble in water. It dissolves in rainwater enriched by carbon dioxide derived from organic soils so the processes and results are on a larger scale in areas of warm, wet climate.
Karst features are erosional forms produced by solution on bare rock surfaces, beneath the soil at rockhead, and within the rock.
Solution is highly selective, so that most joints are etched out to create fissures, gullies and caves; they may be full of air, water or soil, between remnant blades of strong, unweathered rock. This creates the highly variable ground conditions which typify limestone areas.
Pinnacled rockhead describes a highly fissured limestone surface beneath a soil cover. Tall, narrow, unstable or loose pinnacles may be supported only by the soil, and fissures may extend far below into caves. Rockhead relief in tropical areas may be > 20 m.

SINKHOLES

These are any form of closed surface depression with drainage sinking underground. Different types have different implications for engineering activity.

Solution sinkholes develop slowly like blind valleys; slow rate of formation creates no subsidence threat.

Collapse sinkholes are not common, and events of rock failure are rare. Collapse processes contributes to forming many sinkholes; over geological time, they can create zones of broken unstable ground in limestone.

Buried sinkholes provide potential differential settlement over compacting fill. May be conical, cylindrical or irregular; isolated or clustered; 1–50 m deep, 1–200 m wide. Effectively represent an extreme form of rockhead relief with short buried valleys.

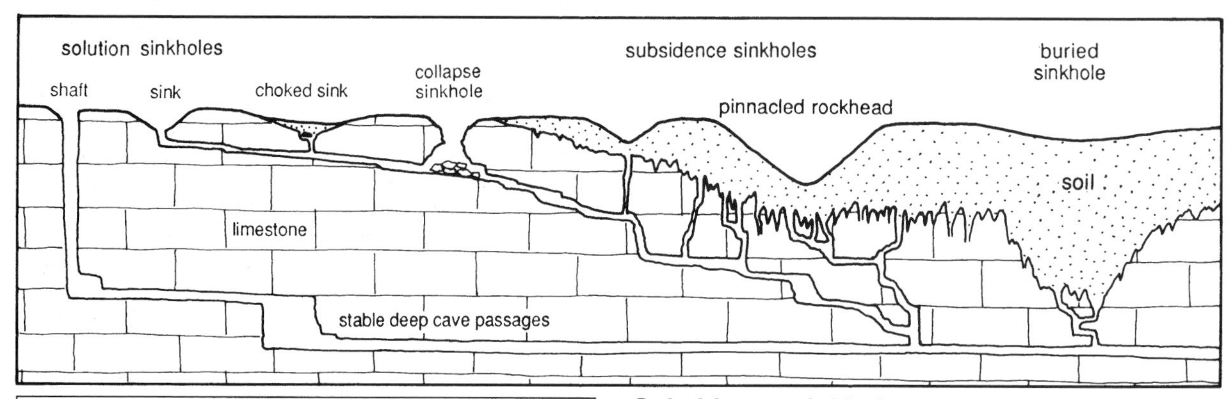

GROUND INVESTIGATION ON LIMESTONE

Many boreholes are needed to map pinnacled rockhead and buried sinkholes, and many rock probes required to prove solid rock without caves (see Remouchamps example in section 23).
Local and site history is the best guide to cave and sinkhole hazard. Shale boundaries and fault lines may have concentrations of sinkholes and caves.
Deep probes should prove bedrock to depth > likely cave width; may need splayed borings to prove that pinnacles are sound.

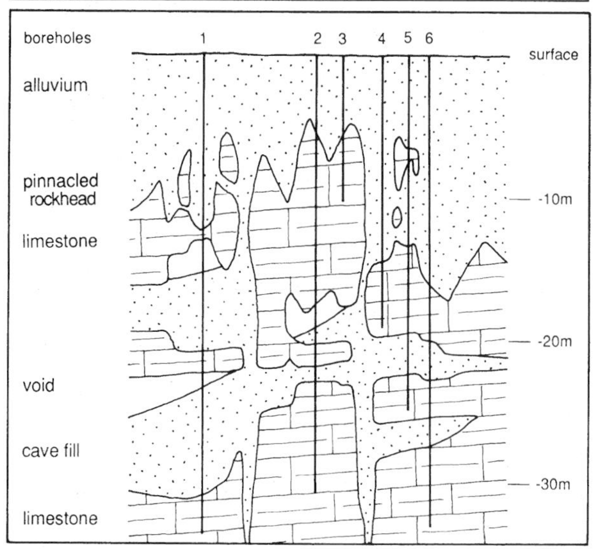

Boreholes on a site in Kuala Lumpur, Malaysia, with an interpreted geological profile reflecting solution along fissures in the massive limestone

Subsidence sinkholes account for 99% of ground collapses on limestone. They form in soil cover, above cavernous rock, due to downwashing of soil (ravelling) into bedrock fissures. Sinkholes may be 1–100 m across. Locations are unpredictable; mostly in soils 2–15 m thick.
In sandy soils surface slowly subsides.
In clay soils cavity forms first at rockhead, then grows in size until cohesive soil bridge fails, to cause sudden dropout collapse of surface.

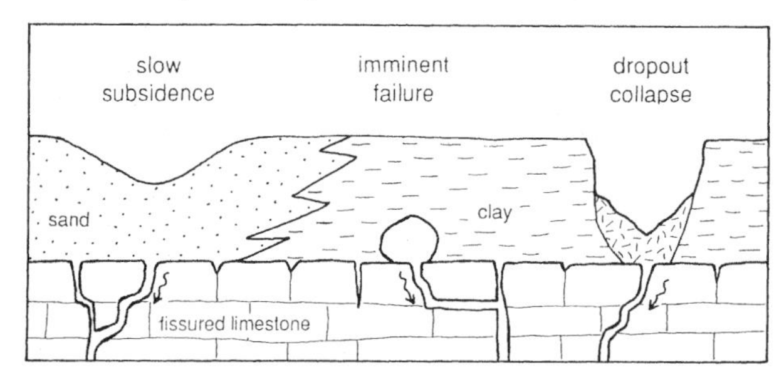

Induced subsidence sinkholes are more common than natural failures; caused when and where drainage through rockhead increases, so washing away more soil; most events are triggered by rainfall.
Water table decline effectively induces sinkholes, mostly when it declines past rockhead; large areas are affected by overpumping for supply, as in Florida, and smaller areas by quarry, mine or site dewatering.
Uncontrolled drainage diversions on construction projects cause many new sinkholes; also structural loading, excavation, devegetation, irrigation and leaking pipelines. Unlined drainage ditches and soakaway drains must be avoided on limestone, especially in alluviated valley floors.

FOUNDATIONS ON LIMESTONE

Driven piles may lose integrity where they bear on rock over a cave, are bent due to meeting a pinnacled rockhead, or are founded on loose blocks or unstable pinnacles within the soil.
Concrete ground beams may be aligned or extended to bear on rock pinnacles which have been proven sound; aggregate pad, stiffened with geogrid, may act in same way and avoid loading the intervening soil.
Can inject stiff compaction grout or polyurethane foam to stiffen soil over limestone, and lift a structure, but a fluid grout injected into limestone can incur large losses into adjacent caves before sealing karstic fissures.
Essential to control drainage over or into soils above limestone, to stop new subsidence sinkholes forming.
Strip or raft foundations can be designed to span any small failures which develop subsequently.
May be best to avoid small limestone outcrops.

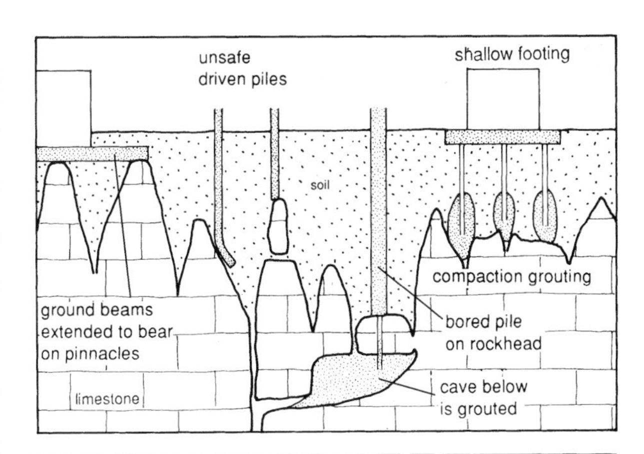

Sinkhole repair must prevent soil entering the bedrock fissure while allowing drainage without diversiôn of water to another unprotected fissure. Coarse rock fill with filters and reinforced soil over is effective. Uncontrolled filling always leads to subsequent renewed failure.

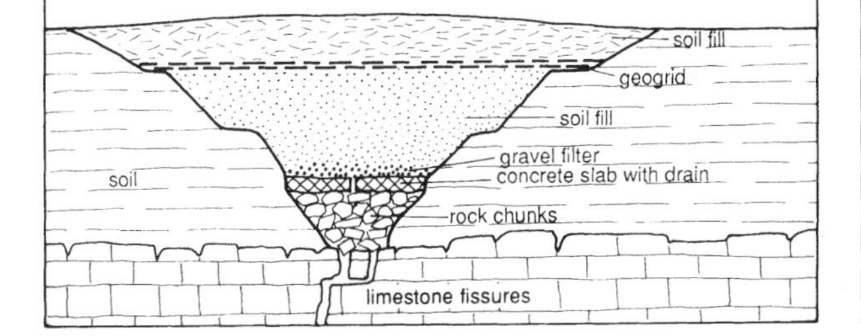

CAVES IN LIMESTONE

Fissures are opened by solution, until they take all available drainage underground, and evolve into an infinite diversity of cave passages and chambers. In many limestones, most caves are < 10 m across, but some tropical areas have cave chambers > 100 m wide. Bedding planes and fractures influence the shape of most cave passages.
Cave locations within a limestone are unpredictable. They commonly have no surface indication; though isolated cavities cannot exist, entrances may lie hidden beneath soil or be only small tortuous fissures.
Cave roof collapse is only likely where the solid cover thickness is less than the cave width but small individual cavities can permit punching failure and threaten the integrity of individual piles or column bases. Statistically, most caves are deep enough to have no direct influence on surface engineering.

FOUNDATIONS ON CHALK

Chalk is weak, friable, pure limestone; when fresh, USC = 5–27 mPa; but porosity is 30–50%, so UCS reduces to 50–70% when saturated.
It may have solution features, caves and sinkholes, but generally on a smaller scale than in strong limestone.
Weathering of chalk by frost action is severe, to produce weak rubbly debris. This commonly reaches depth of 10 m in Britain due to periglacial weathering during the Ice Ages.
Putty chalk and fine grained rubble chalk are thixotropic when saturated, and turn into slurry when disturbed. Should not be excavated or handled in wet winter months, but can be used as fill when dry.

Pile driving in chalk creates slurry at tip; this stabilizes when left undisturbed, so piles may carry higher working load if left for a time after driving. Settlements in chalk are often lower than expected, as rock strength increases under steady load.
Driven concrete piles have ultimate end resistance of N/4 MPa, where N = SPT count. Risk of solution cavities below pile tip means that load is best attained by shaft resistance with ultimate values of 30 kPa on displacement piles and 150 kPa on cast-in-place piles.

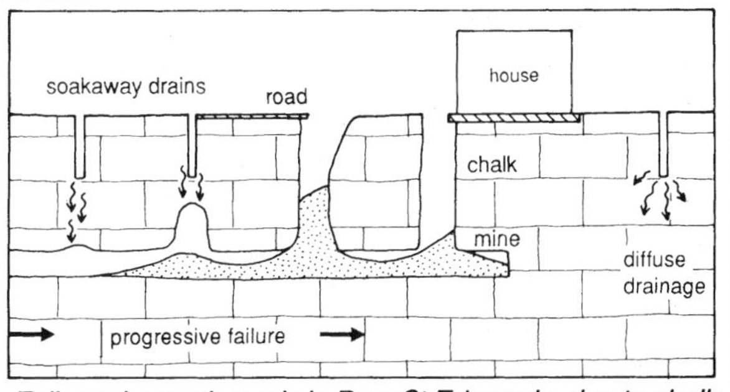

Failures beneath roads in Bury St Edmunds, due to chalk liquefaction between soakaway drains and old mines.

Liquefaction failure of putty chalk occurs where it is saturated along route of concentrated drainage and can fail into a cavity beneath, usually a mine, gull or cave.
Ground collapses at Norwich and Bury St Edmunds (in East Anglia) and at Reading are mostly related to old mines below soakaways or drain failures; some are collapses of clay-filled pipes within the chalk.
Good surface drainage and ban on soakaways are necessary in chalk areas, especially where voids may exist – where there is a history of mining, along cambered scarps with gulls, or on valley floors underlain by caves.

Chalk properties relate to grade of weathering. Tabulated values are typical for Middle Chalk. Porous Upper Chalk is commonly weaker.

Weathering grade	Description	Creep at 400 kPa	SPT N	SBP kPa
V	structureless putty	significant	< 15	50–125
IV	friable rubble	significant	15–20	125–250
III	blocky rubble	small	20–25	250–500
II	medium hard	negligible	25–35	500–1000
I	hard and brittle	negligible	> 35	> 1000

30 Subsidence over Old Mines

Ground stability ultimately depends on the style of mining utilized, which is generally dictated by the shape, size, depth and value of the ore or extractable rock.

STOPING Conventional deep mining, of mineral vein or any shape of orebody, creates large open underground voids known as stopes. Subsidence threat is localized, but may totally sterilize narrow strips of ground directly over the mines; a wider potential hazard is failure of hanging walls left above inclined stopes.

PILLAR AND STALL Deep mining of thin low-dip beds by partial extraction; utilized for most coal working before 1940. Between 10% and 40% of ore is left in place to form pillars to support roof – in random plan or systematic rooms, stalls or bords in old hand-worked workings, or in regular grid in modern mines. Older mines, often over-extracted, create a long-term subsidence threat, but better controlled modern mines have no surface effect.

LONGWALL Total extraction of coal in modern mines, with automatic, immediate, surface subsidence – section 31.

SOLUTION Wild brining pumps natural brines from salt beds at shallow depth, and greatly accelerates linear subsidences above natural brine flows (section 27). Controlled brining pumps fresh water into, and brine out of, salt at depth, and should be totally stable.

OPEN PIT or QUARRY Total extraction of bulk rock (quarry) or mineral ore (open pit) together with any waste rock needed to ensure pit wall stability. Backfill is rarely possible or economic in large workings, except for some waste rock fill in worked out areas. Small old quarries are far more numerous, often with unstable and compressible fill of domestic refuse.

OPENCAST Continuous operation of surface excavation, ore removal and backfilling with displaced overburden. Draglines may cast the overburden over the site of ore removal, or earthscrapers are used to take it around the site. Approximate ground level is restored with uncompacted fill which is graded to desired profile and re-covered with topsoil; bulking of broken fill roughly compensates for ore removal. Commonly used for modern coal working; multiple seam extraction may leave benched rockhead profile beneath fill.

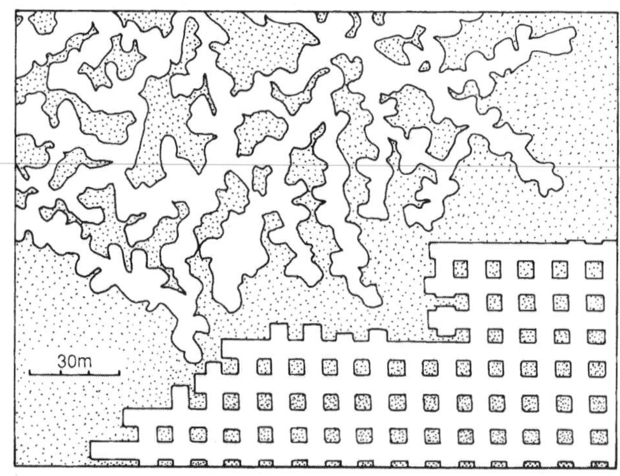

Pillar and stall mining of gypsum bed 2 m thick.
Above: plan of old irregular and modern regular workings.
Below: stable mudstone roof in the modern workings.

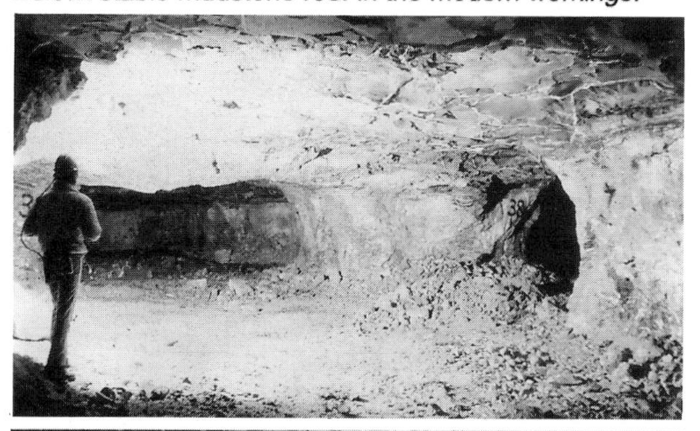

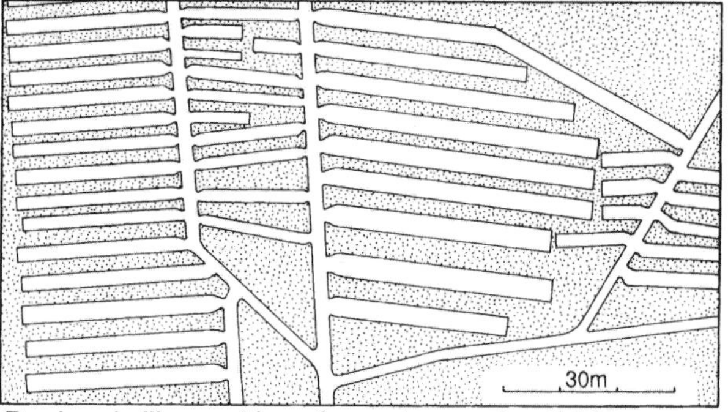

Bord and pillar working of a coal seam 1 m thick.

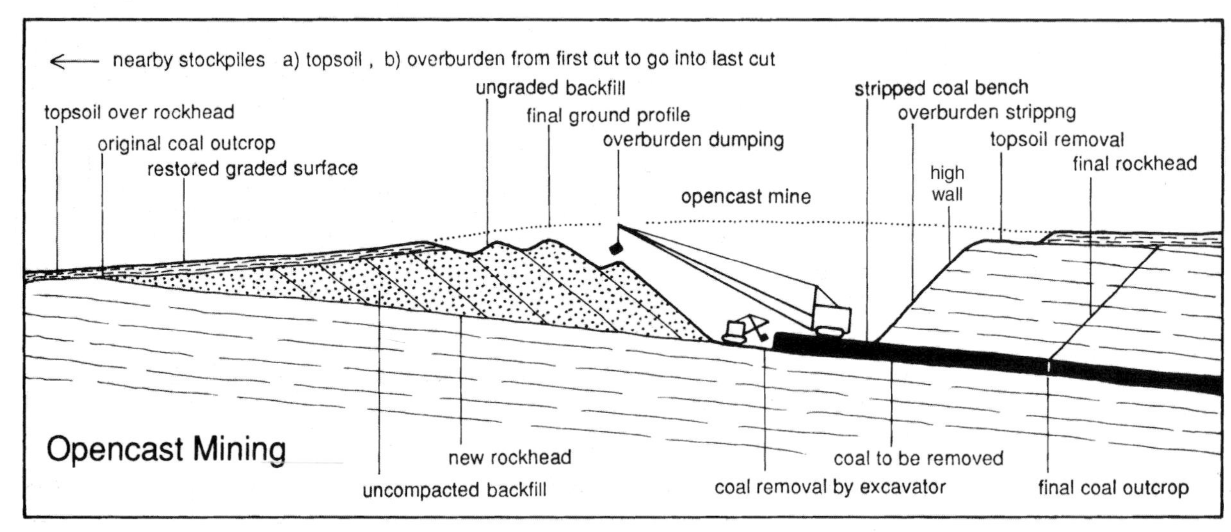

ROOF FAILURE AND CROWN HOLES

Roof span failure and progressive breakdown of beds causes upward stoping (migration of cavities).
This may reach the surface to create a crown hole by sudden collapse, or stoping may be stopped by beam action of a strong bed, by formation of a stable arch in thinner beds, or by bulked breakdown debris meeting and supporting the roof.
Crown holes are rare from mines at depths greater than about 30 m or 10 times extracted seam thickness.

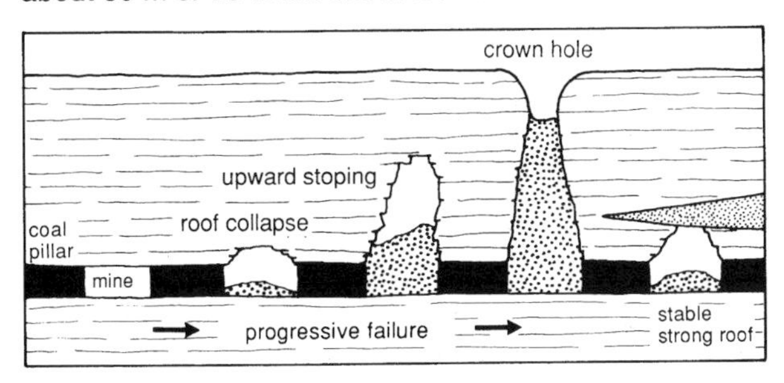

FAILURE OF OLD SHAFTS

Thousands of old mine shafts are a widespread hazard. Small old mines had far more shafts than large modern mines; records of old shafts are very incomplete, and site investigation must pursue any documented or physical indication.
Shafts are mostly 1–5 m diameter, 10–300 m deep; may be lined with brick, concrete or dry stone or may be unlined in rock; may have loose or compacted fill to bottom of shaft or above unstable stopping, or may be empty; may be covered with timber, vegetation, steel or concrete, or may be well sealed and capped.

Shaft cap should be reinforced concrete slab of diameter 2·6 times that of shaft, founded on sound rock. Coal measure shafts usually require filling, or grout injection of old fill, for development within 20 m; remedial costs may exceed £10 000.

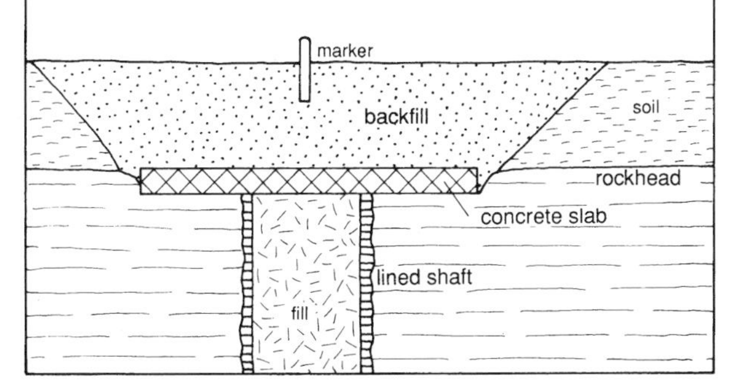

SAFE COVER FOR OLD MINES

Guideline figure is 30 m for old coal mines, so this is also the minimum depth for borehole investigation.
At > 30 m depth, pillar failure is rare because imposed load is small compared to overburden load, and roof stoping rarely reaches the surface to form crown hole.
Even within coal measures, local conditions may vary, with strong sandstone roof or old weak pillars eroded by water; some mines 10 m down are stable for houses; others have needed filling at 50 m depth.
Safe depths are different for rocks other than coal measures; buildings are safe 3–5 m above old mines in Nottingham sandstone; pillar failure in limestone mines 145 m down near Walsall caused surface subsidence after stoping collapse of mainly shale cover.
Remedial costs may exceed £50 000 per hectare, but should be < 5% of project costs; shaft filling or major grouting exercises incur maximum costs.

MINE PILLAR FAILURE

Mine pillars fail where they are left too slim, are subsequently overloaded or are subject to weathering and erosion. Multiple failures, domino-style, may affect large areas and were common in the past due to over-extraction and pillar-robbing.
Collapse of old mines may be delayed 100 years or more. Modern threat of ground failure is minimal where mine is > 50 m deep, where any imposed structural load is slight in proportion to existing rock overburden load and where pillar erosion is generally less than it is near to the surface.

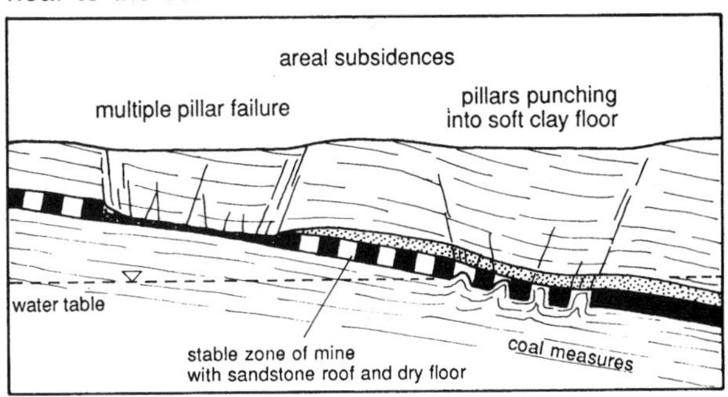

TREATMENT OF OLD MINES

Excavation and backfill is normally only feasible and economic to < 5 m depth.
Piling is normally limited to 30 m depth, and only through drift or shale, as boring through sandstone is uneconomic; cannot be used where dip is steep, where there is any risk of sliding, or where deep mining subsidence is active or anticipated.
Grouting may need 100 mm bored holes on 3–6 m grid to ensure complete filling. Must include marginal zone of width that is 0·7 times depth to encompass zone of influence. Perimeter is sealed first; grout stiffened with pea gravel forms cones around holes bored on 1·5 m centres, that coalesce to create a wall within the mine.
Can fill with low-strength foamed concrete or lean rock paste to prevent roof collapse between sound pillars.
Founding on rafts or reinforced strip footings may be good for low-rise buildings over mines of marginal depth, where risk does not warrant expense of filling.

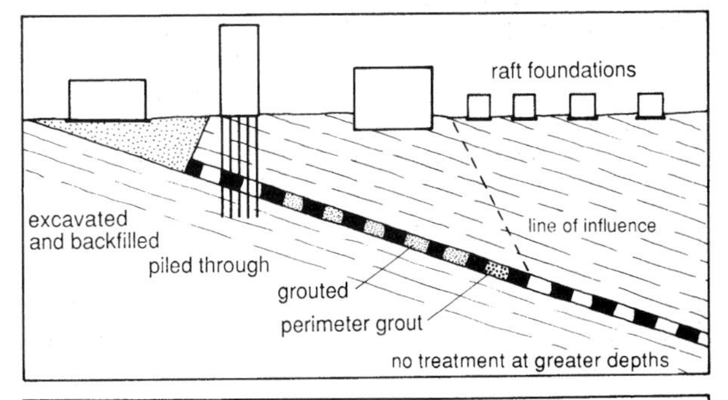

Bell pits are shafts usually < 10 m deep to old coal workings reaching only a few metres from the shaft and not interconnected. Generally occur in dense groups, and must be filled or excavated if development over them cannot be avoided.

31 Mining Subsidence

Total extraction mining removes all the mineral from a bed, allowing the unsupported roof to fail and cause inevitable and predictable surface subsidence. It is used worldwide in a large proportion of modern underground coal mines.

Longwall mining is the method used in Britain. Extraction is by a machine coal cutter moving back and forth along a single migrating coal face up to 400 m long. After a slice of coal about 1 m thick has been cut from the whole length of the face, the hydraulic roof supports are advanced, the roof behind is allowed to fail, and the process is repeated.

Panel of coal is removed; about 300 m wide and maybe over 2000 m long, with no support beyond the working face and access roadways.

Alternative method is a version of pillar and stall mining followed by pillar removal on the retreat. Surface subsidence effect is same as for longwall.

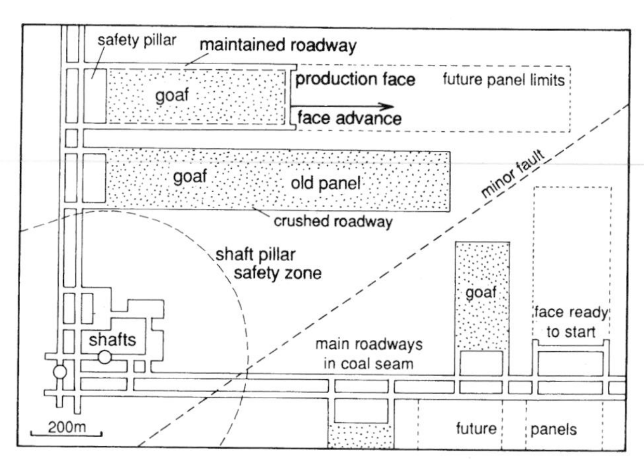

Layout of panels in a typical modern coal mine

SURFACE SUBSIDENCE

The ground surface is deformed above a working coal face by a subsidence wave which migrates at the same rate as the face advance – usually 10–20 m per week. This subsidence wave has a number of effects:

Subsidence must be less than seam thickness, so usually about 1 m; may accumulate to > 15 m by multiple seam working over time; causes little structural damage but has impact on drainage and piped services.

Ground strain develops first as extension (on the convex part of the wave), then a return to neutral, followed by compression (on the concave part of the wave). This causes most of the structural damage due to mining subsidence. Total strain is the sum of extension and compression values and is typically 1–10 mm/m or 0·001–0·01.

Angular movement occurs as tilt on the subsidence wave; usually minor and only significant to tall chimneys and sensitive machines.

Micro-earthquakes may occur due to movements in strong, massively jointed rocks under stress.

At any one site, subsidence movements are generally completed within no more than one year.

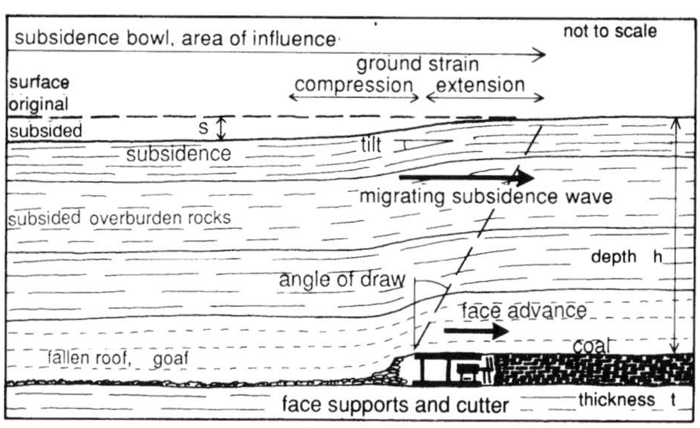

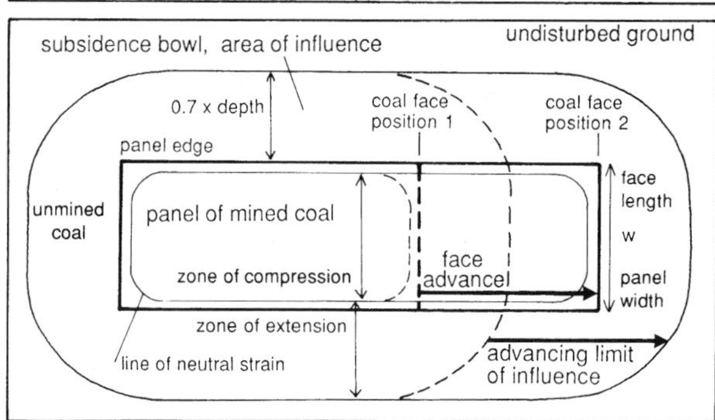

PATTERNS OF SUBSIDENCE

Mining subsidence follows a well defined pattern.

Depth and lateral extents of the subsidence bowls and strain profiles, can be predicted on the basis of many past measurements, and the empirical data conforms closely with theoretical calculations.

Critical parameters which determine subsidence movements are the depth of working (h), the panel width (w), and the extracted thickness of coal (t).

Above an extracted panel, the ground moves downwards and also inwards, so that an area of ground larger than the panel is affected. The angle of draw is normally 30–35°, increasing slightly in weaker rocks.

Area of influence extends 0·7h outside the panel; edge is not clearly defined as it tapers to nothing.

Subsidence wave has a length of about 1.4h, with a midpoint of maximum tilt and neutral strain, close to vertically above the coal face. It migrates with the advancing face and also develops to a similar shape over the panel sides.

At any one point on the surface, movement occurs over the time taken for the wave to pass, typically 38 weeks for a 560 m long wave over a 400 m deep face advancing at 15 m/week.

Strain profiles show outer zone of extension and inner zone of compression; line of neutral strain is roughly above panel edge, varying slightly with changing w/h ratio.

Maximum strains are close to panel edge. Residual compression falls to zero over centre of panel where w/h > 1·4.

Subsidence and strain are most severe over shallow, wide panels in thick seams; they are also complicated by geological factors (faults, strong rocks, steep dips) and multiple workings.

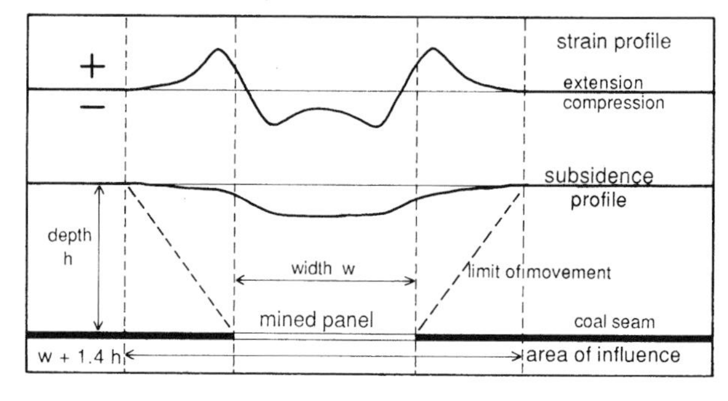

PREDICTION CALCULATIONS

Approximate predictions can be read from graph on the right which shows maximum values of subsidence strain and tilt related to h, w and t. These are typical values only; better predictions are made with graphs for specific coalfields based on their own records and rock characteristics.

This graph only gives maximum values; partial strain and subsidence, which occur outside side edges of panels can be read off more detailed graphs.

Example of calculations using this graph:

Site factors (from mine plans): thickness = t = 1·2 m; panel width = w = 160 m; depth = h = 400 m.

Ratios: w/h = 160/400 = 0·4; t/h = 1·2/400 = 0·003.

- Reading off graph for value of w/h = 0·4:

Subsidence factor (direct from graph) = s/t = 0·3

Subsidence = s = 0·3 × t = 0·3 x 1·2 = 0·36 m = 360 mm.

Extension = E = 0·28 (from graph) × t/h
= 0·28 × 0·003 = 0·00084.

Compression = C = 0·62 (from graph) × t/h
= 0·62 × 0·003 = 0·00186.

Strain = E + C = 0·00084 + 0·00186 = 0·0027 = 2·7 mm/m.

Tilt = 1·4 (from graph) × t/h
= 1·4 × 0·003 = 0·0042 = 1 in 238.

CONSTRUCTION IN SUBSIDENCE AREAS

Concrete rafts are simplest and cheapest foundations for buildings; smooth based, formed on polythene over 150 mm granular sand to absorb horizontal strain; reinforced both top and base, maximum 20 m long or with stiffening beams on top.

Structural units should be as small as possible, or may be articulated to tolerate strain.

Alternatives are deformable structures, some with sliding panels and spring bracing.

Piles need care as tilting can diminish integrity.

Pipelines need flexible joints, and gravity flow drains need slope greater than predicted tilt.

Bridge decks may be on three-point roller or spherical bearings, with hinged piers, and bitumen or comb expansion joints.

Jacking points for bridges, machines or buildings are cheaper built-in than added later.

PRECAUTIONS FOR OLD STRUCTURES

Most structural damage is under tensile strain; tiebars can be added to buildings.

Ground compression can be halved by digging trenches around a building to isolate a raft of soil beneath it.

Bridges may need temporary support or deck removal.

Pipelines can be exposed and placed on sliding chocks.

SUBSIDENCE COMPENSATION

Longwall mining has compensation for inevitable subsidence damage incorporated in its budget, but compensation law varies between countries.

In Britain, Coal Authority pays costs of damage repairs, except to recent structures where precautionary measures were appropriate but were not installed.

Coal Authority does not pay cost of precautionary works, even if these are required by local planning law.

Subsidence may be reduced, but not eliminated, by stowing waste before roof collapse, by leaving support pillars, or by harmonious working (where one panel's compression cancels out a second panel's extension).

Urban areas, where compensation costs may exceed 20% of coal's value, are not now undermined in Britain.

When mining in a region ceases, drainage pumps are switched off; then groundwater rebound raises joint water pressures and re-activates over-stressed faults; may cause new phase of localized ground movement.

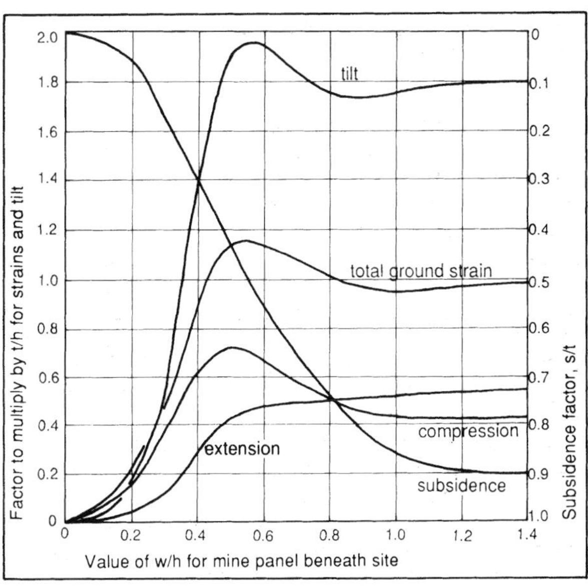

STRUCTURAL DAMAGE RELATED TO STRAIN

Damage relates to both ground strain and structural size.

Potential distortion = strain × structural strength

Class of damage and distortion	Typical features of damage
Very slight < 30 mm	Barely noticeable hair cracks in plaster
Slight 30–60 mm	Slight internal fractures, doors and windows may stick
Appreciable 60–125 mm	Slight external fractures, service pipes may fracture
Severe 125–200 mm	Floors slope and walls lean, doors frames distorted
Very severe > 200 mm	Severe floor slopes and wall bulges, floor and roof beams lose bearing, needs partial or complete rebuilding

GEOLOGICAL FACTORS

Some ground conditions create very variable subsidence and make detailed site predictions very difficult; these geological factors account for 25% of movements and damage being outside predictions, either above or below.

Fractures (joints and gulls) in strong, competent rock at outcrop localize movement, creating zones of very high strains between stable areas where blocks of rock act as natural rafts. Sandstones, and the Magnesian Limestone of northern England, develop open fissures under tension, with subsidence sinkholes in soil cover.

Faults localize movement with zones of high strain and ground steps due to displacement.

Steep dips displace the subsidence bowl in downdip direction and significantly distort strain profiles.

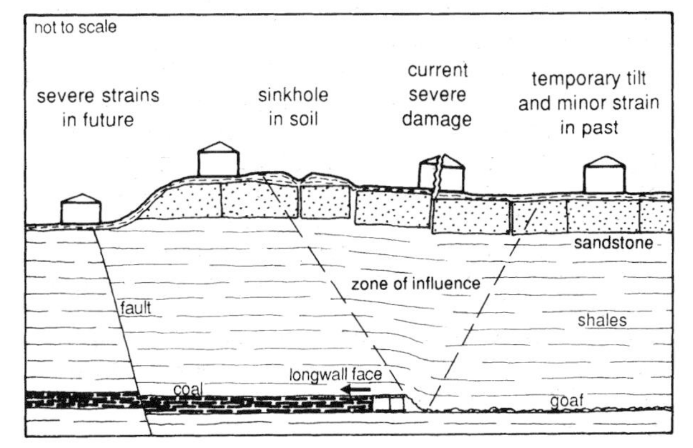

32 Slope Failure and Landslides

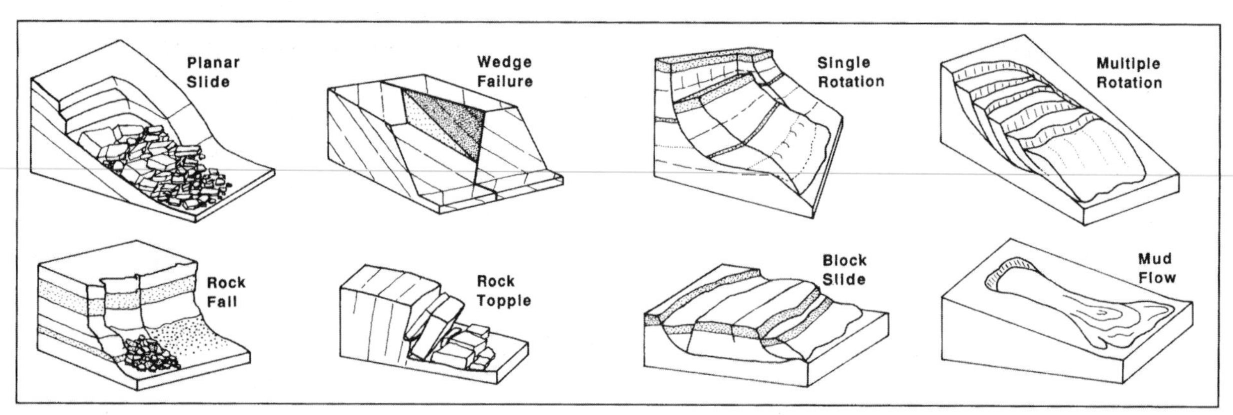

Nearly all slopes ultimately degrade by the natural processes of weathering and downslope by transport.
On most slopes this is a continuous, very slow process.
Landslides occur where a slope remains static for a long period and then fails in a single dramatic event.
The result in both cases is the same; landslides are one end of a spectrum of natural processes.
Landslides may occur in any rock type.
They commonly occur where some geological structure, weakness or contrast interrupts the pattern of slow degradation.
Potential landslide sites are recognizable by their geological structure.
Each landslide is normally triggered by an individual event or process.
Landslides are only understood by assessment of both the initial structure and the trigger process.

UNSTABLE SLOPES

Each rock material has its own equilibrium slope angle.
Clays are generally unstable at >10°, roughly $\phi/2$.
Most rocks of moderate or greater strength can be stable in vertical walls 100 m high if they are massive with only vertical and horizontal fractures. Granite forms a vertical wall 700 m high on Half Dome California, and the vertical cliffs 150 m high at Beachy Head, Sussex, are formed in much weaker chalk. Minor rockfall is always a hazard on these high faces.
Planar weaknesses – bedding planes, joints, etc. – inclined towards the slope create potential slip surfaces in any rock; slopes degrade back to any major fractures with dip $> \phi$ (may be < 20° for clay infilling; cohesion and water pressure are also significant).
Densely fractured or thin bedded rocks weather back to slopes of 20–40°.
Potential failure can be assessed on any of the above criteria in context of local data. Rock slides are mostly related to bedding planes, joints, faults, cleavage or schistosity planes which daylight (have unfavourable orientation and are exposed at their lower end) in a surface slope.

TYPES OF FAILURE

Large rock failures are mostly planar or wedge slides on one or more plane surfaces.
Small rock failures are commonly falls or topples.
Clay failures are mainly single or multiple rotational slides, ideally on circular slip surfaces.
Mud slides, mud flows and debris flows develop from weak clays or in failed rock material after initial displacement.
Complex failures are common and involve multiple processes; block slide has planar, and circular head.

Head scar of a rotational slide breaks a road in Yorkshire

SPEED OF FAILURE

Slow: more common in soft clays and ductile materials, notably reactivated old landslides. Thistle Slide, Utah, 1983, moved < 1 m/h for two weeks.
Rapid: typical of brittle rock failures as rock is greatly weakened by initial shearing or fracturing. Velocities of > 100 km/h are common, as at Madison Canyon.
Cyclic: failure creates a stable slope as the debris becomes toe weight, but erosion of the debris then permits repeat failures, as at the frequent landslides along the boulder clay cliffs of the Humberside coast. Alternatively, due to annual changes of groundwater levels; Mam Tor Slide (section 35) moves every winter but is stable in summer.

MADISON CANYON LANDSLIDE, MONTANA, USA

Geology and slope angle varies along canyon wall.
West part: 45° slope in strong dolomite – stable.
East part: 30° slope in weak schist – stable.
Mid part: dolomite buttress below schist slope – unstable.
Increased stress in earthquake broke dolomite buttress; unsupported schist slope then failed: 20M m^3 landslide.
Failure of this part of the slope was inevitable, whenever dolomite buttress was adequately eroded or weakened; vibration from earthquake was just the trigger process.

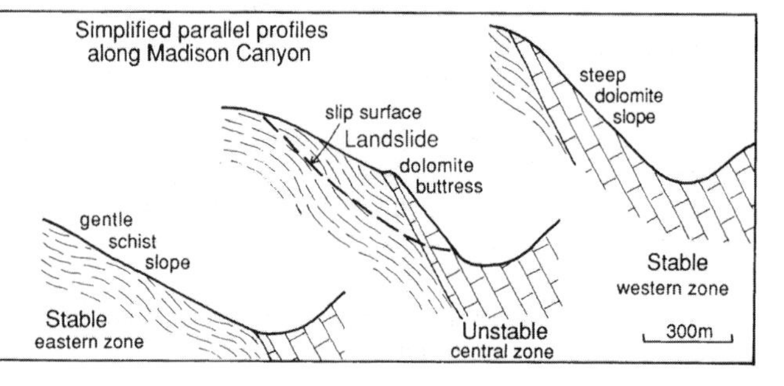

GROS VENTRE SLIDE, WYOMING, 1925

Thick strong sandstone above bed of weak clay, all dipping 18° towards river valley.
River eroded toe of slope, removing sandstone support, until 38M m^3 moved in bedding plane slide.
Similar to prehistoric slides on same side of valley; opposite steeper slope is stable.
Debris blocked valley, creating new lake; first overtopping by river eroded debris and caused downstream flood.

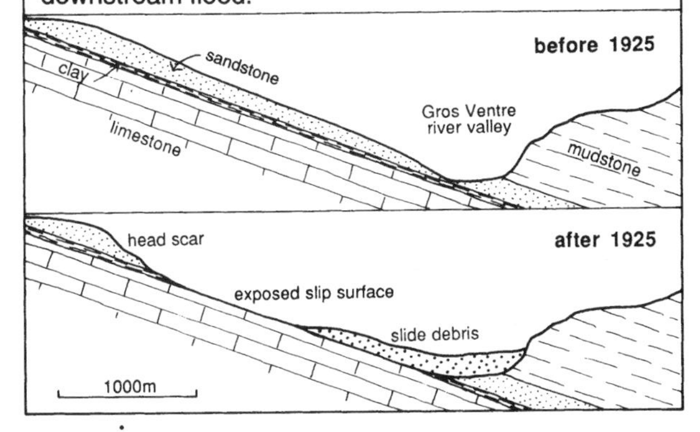

LANDSLIDE TRIGGER PROCESSES

Each landslide event can be ascribed to a process which triggered the failure of a potentially unstable rock mass.
Cause of failure is therefore a combination of unstable structure and a trigger event.

Water: rise in groundwater pressure is by far the most important single trigger factor behind landslides – see section 33.

Toe removal: removing toe of a slope reduces resistance to movement.
Natural toe removal: erosion by river undercutting (Gros Ventre, above); erosion by wave action causing numerous coastal slides (Folkestone Warren, section 36); glacial erosion leaving oversteepened hillsides (Mam Tor, section 35).
Artificial toe removal: by quarrying or mining (Frank), excavation for building site (Hong Kong), or road widening (Catak) (all in section 33).

Head Loading: adding material above neutral line of a slide increases its driving force. Portugese Bend slide, Los Angeles, 1956, activated by fill placed for a new road which added 3% to slide mass in zone above slip surface dipping 22° in weak clays. Folkestone slide, 1915, activated by rock falls from head scar (section 36).
Natural head loading causes slope instability on many active volcanoes.

Strength reduction: weathering ultimately weakens all slope materials; slow creep causes restructuring of clays stressed within slopes (section 34); slow processes eventually reach critical points.

Vibration: cyclic and temporarily increased stresses may cause soil restructuring or rock fracturing.
Artificial vibration, as from heavy road traffic (contributory at many small road failures) or from pile driving (which caused a clay slide destroying Swedish village of Surte in 1950).
Earthquake vibration has caused numerous slides. 1970 earthquake in Peru started slide on Mt Huascaran which developed into debris flow moving fast enough to rise 150 m over ridge and bury 20 000 people in the town of Yungay.

Many slides have complex origins, where and when a number of contributory factors coincide.

STABILITY OF A SLIDE MASS

Basic forces on a slide block are:
W = weight of block, with two components, D and N.
D = driving force = W sin α.
N = normal stress on slip plane = W cos α – u.
u = uplift force due to pore water pressure.
c and F = resistances in reaction to D.
c = cohesion across slip plane.
F = frictional resistance on slip plane = N tan ϕ
R = resistance to shear = c + (W cos α – u) tan ϕ
Safety factor = R / D = resistance/driving force.
c and ϕ are properties of the rock material.

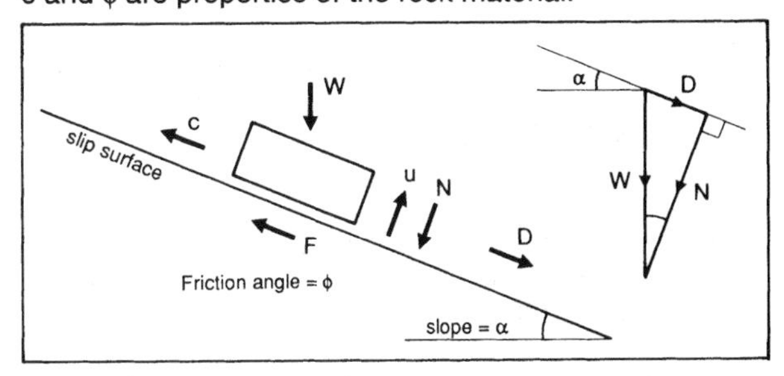

Neutral line: a curved slip surface beneath a slide mass has a neutral line boundary (NL) between a steep section where D > R and a flatter section where R > D.
Tension joints or open fissures at head of slide may contain water exerting a horizontal joint water pressure (J) which adds to the driving force.
Deformation within the slide mass must occur as it moves over a slip surface which is other than plane or cylindrical; resistance to this deformation by cohesion and friction along multiple internal slip surfaces adds to the resisting force.

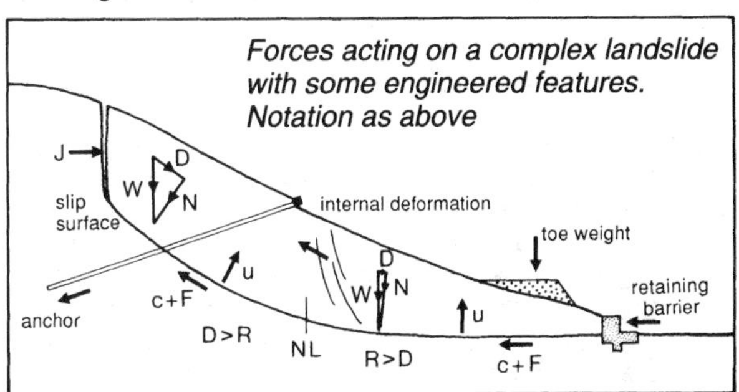

Forces acting on a complex landslide with some engineered features. Notation as above

Stability analysis of a landslide may be by assessment of forces in two dimensions in individual slices of the mass; these vary across the slide and may include artificial constraints.
Full landslide stability analysis is more complex due to:

- breaking slide into small units
- reaction forces between these units
- variable water pressures
- estimated values of c and ϕ
- reactions in three dimensions.

Force diagrams drawn to scale quantitatively represent components in a two-dimensional slope stability analysis.

sliding force F Ac α ϕ Weight u W J u F+Ac area A α

slice width = 1m units = kN Joint water pressure

33 Water in Landslides

Groundwater is the most important single factor in triggering landslide events.
Rising water tables and rising water pressures, are contributory to most slope failures; the majority of landslides occur during rainstorms.
Effective stress is reduced by any increase in water pressure and there is a consequent reduction in resistance to shear (sections 26 and 32).
Joint water pressure in rocks and pore water pressure in soils are equally important.
Drainage is therefore critical to slope stabilization (section 36).
Water input to a slide mass also has the long-term effect of internal weathering. Loading by water in a slide mass may increase the driving force.
Water does not act as a lubricant. The only material approaching the properties of a lubricant in a slide is clay softened by increased water content.

SCARBOROUGH LANDSLIDE, 1993

Single rotational slide formed in 30 m high clay slope, then retrogressive failure of head scar destroyed hotel. Long period of drought had left shrinkage cracks in the dried clay, followed by heavy rainfall which raised pore water pressure.

FRANK SLIDE, CANADA, 1903

Glacially over-steepened slope in Rockies cut in dipping limestone of marginal stability.
Rock fissures opened due to creep movement initiated by crushing of inadequate mine pillars left in vertical coal seam across toe of slope.
Failure occurred after first day of spring thaw.
Rockfall of 37M m^3 buried mining town of Frank.
Cause was a combination of the dipping limestone, the mined toe and the snowmelt input.

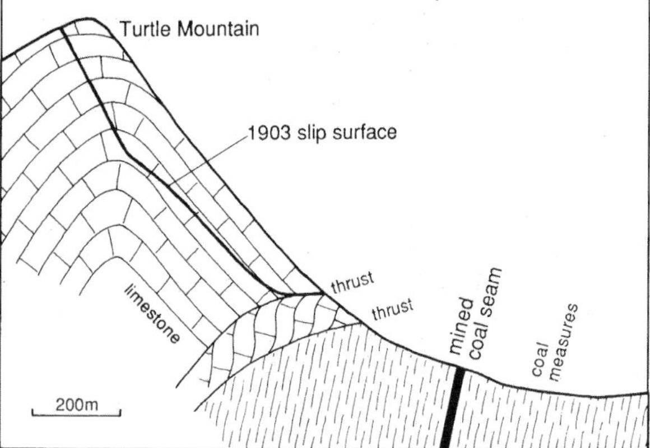

SOURCES OF INPUT WATER

Rainstorms: High rainfalls from individual storms cause numerous shallow slides where high water pressures can rapidly reach slip surfaces.
Hundreds of slides in Jordan, in early 1992, due to rare heavy snowfall and rapid melt in a normally semi-desert terrain; soils, rocks and fills all equally affected.
Destructive 1988 slide at Catek, Turkey, failed during first period of high rainfall since road widening had steepened the slope four years previously.
Shallow earth slides are annual events during rainstorms on steep slopes of the shanty town favellas in Rio de Janeiro.

Rainfall seasons: Deep-seated slides are more affected by annual fluctuations of water table.
Winter groundwater maxima create landslide season from November to March in Britain.
Monsoons cause most of the landslides in SE Asia.
Spring snowmelt is the main factor in alpine regions after slopes have been stable during winter freeze.
Numerous sets of data show correlation between rainfall and slide movement; mostly on small scale with rapid response, or on large scale with response delayed 1–10 weeks; Portuguese Bend and Vaiont are examples.

Artificial inputs: Impounding water in a reservoir raises regional water tables, as at Vaiont.
Devegetation of a slope allows increased infiltration.
Irrigation of farmland or gardens has caused many terrace edge failures in dry regions such as California.

Secondary effects: Opening of tension fissures in head zone, as slide starts to move, captures runoff and increases infiltration.

PORTUGUESE BEND SLIDE, LOS ANGELES

Coastal slide of 100 ha of weathered shale, on 6° slope.
Slow creeping movement totalling 40 m since 1956 has destroyed 127 houses and damaged the coast road.
Complex relatively shallow slide responds to both individual rainstorms and seasonal weather patterns.

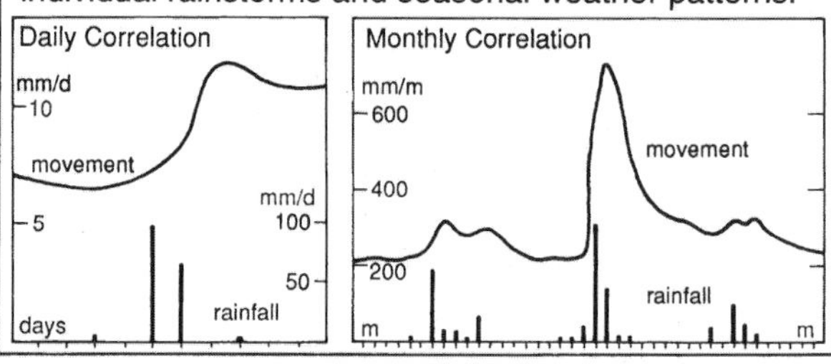

EARTH SLIDES IN HONG KONG

Frequent shallow slides on steep slopes in urban areas.
Slides form in soil layers and decomposed granite and volcanic bedrock (weathering grade IV and V).
Soils drain quickly on steep slopes, but are saturated by periodic heavy rainfalls.
Weathering yields sand and clay soils from bedrock; slope movement creates layered soils.
Sand soils over clay soils are freely drained; recognized as low hazard zones.
Clay soils over sand soils can create inclined artesian aquifers, with high water pressures on underside of clay cap where slip planes can develop; designated hazard zones on steep slopes.
Clay layers are more important as impermeable seals than as easily sheared planes.

VAIONT RESERVOIR SLIDE, 1963

The Vaiont slide involved a magnificent dam, an awful reservoir site, and the world's worst disaster caused by civil engineering with 2043 people dead.

Vaiont Dam: in Alps north of Venice; cupola (double arch) dam 266 m high, of concrete 4–23 m thick.

Slide on 9 October 1963: 270M m^3 of rock, forming a slab 200 m thick, moved 400 m at 20–30 m/s. Landed in reservoir and created huge waves.

Wave 100 m high overtopped dam (which survived); Longarone and other villages destroyed.

Limestones, strong and impure, form slide mass; thin bedded with many clay horizons in lower part; interbed horizons are 5–100 mm thick, plastic clay, PI = 30–60, ϕ = 8–10°; below slide are pure karstic limestones.

Dip = 30–45° N (downslope) at slide head, 10–15° E near valley floor.

Slide mass was massive wedge on bedding plane slip surfaces and along faults on eastern edge. Moved as single slab. It was a preglacial landslide mass, reactivated because new movement was possible into post-glacial Vaiont river gorge through old slide toe.

Groundwater pressures were raised by impounding reservoir; also rose due to rainfall; high pressures beneath slide basal clays, in limestone fed by karst sinkholes high to south. Heavy rain just before failure.

Movement of hillside monitored since dam finished 1960; slip of 0–35 mm/day correlated with discontinuous reservoir filling; also correlated with rainfall in previous 60 days. Small part of slide failed in 1960.

Slip surface largely followed postglacial slip in clay beds; also broke across some limestone beds.

Resistance to shear mainly on eastern side of wedge, ϕ = 36° along fractures.

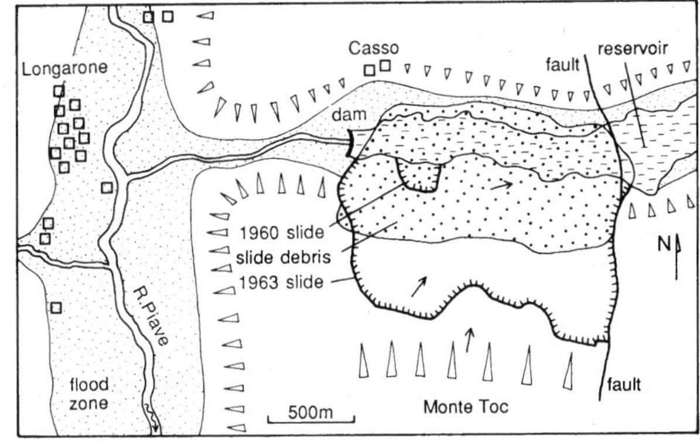

Stability analysis suggests factors of safety (SF).

Reservoir level =	none	Rainfall = low	SF = 1·21
"	none	high	1·12
"	710 m	low	1·10
"	710 m	high	1·00
"	722 m	low	1·00

Reservoir designed to fill to 722 m; failed at 701 m in wet period; would have failed at 722 m in dry weather.

Cause: unstable dipping limestone forming old slide.

Triggers: high rainfall and reservoir impoundment.

Rapid failure: due to brittle rupture of some key limestone beds and rock units, after years of creep had reduced mass strength; borehole monitoring data suggests lack of movement and stress accumulation in toe of slide while surface was creeping.

The error was to assume slide would creep until it stabilized on flatter toe. Potential instability was recognizable; reservoir was inappropriate.

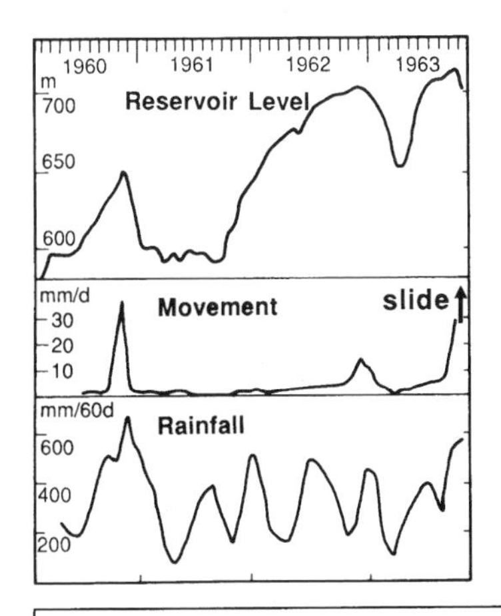

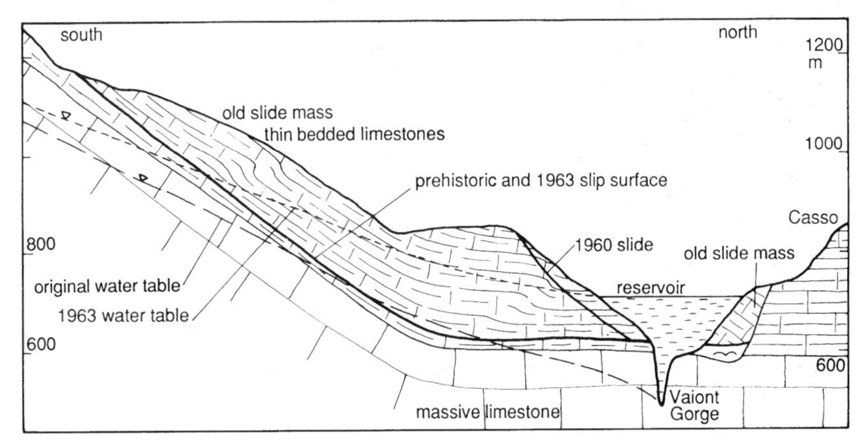

PO SHAN ROAD SLIDE, 1972

The largest slide on Hong Kong island, killed 67 people.
Natural slope at 36°, with soil 15 m thick.
Existing buildings on 20 m piles to sound rock.
Cause: steep soil slope + high rainfall + cut face.

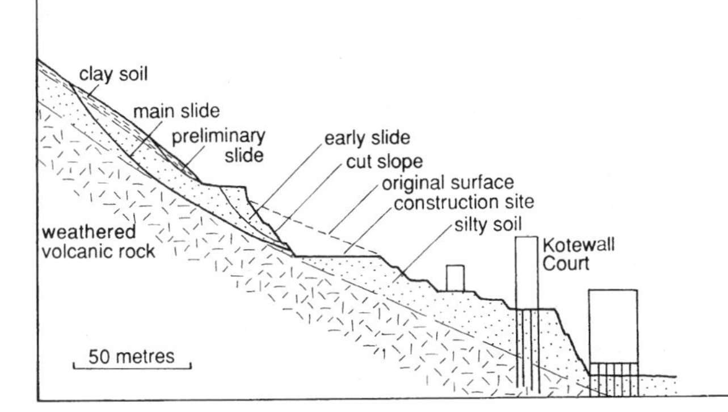

Sequence of events is significant:

1. Construction site left cut slope unsupported for 7 years; exposed soil softened in bare slope at 75°.
2. Small falls, groundwater seepages and head cracks in Po Shan Road observed for a year before event.
3. Rainfall of 700 mm in 3 days prior to slide, highest on record; caused small adjacent slides.
4. The day before main slide, slip produced 5 m head scar in Po Shan Road and debris flows on building site.
5. Four hours before main slide, most of cut face failed and debris flows crossed Conduit Road 2 m deep.
6. Immediately before main slide, small slip from above Po Shan Road landed onto road.
7. Main slide involved 25 000 m^3 of debris, and moved 280 m, in < 1 minute.
8. Kotewall Court, 13 storeys high, pushed off foundations, tilted and collapsed, hitting next block of flats downslope at side of slide.

34 Soil Failures and Flowslides

CLAY SLOPES

Clays are the weakest, most unstable, slope material.
Undisturbed clays can stand in steep temporary slopes held by cohesion, pore water suction and peak frictional strength.
Disturbances or restructuring through creep over time causes realignment of the clay particles to parallel. This reduces internal friction and eliminates cohesion; at the same time drainage equilibriates and eliminates suction.
Natural slopes, with long-term stability, depend on internal friction only. Saturated clay soils have nearly half their weight carried by pore water pressure; so slope is stable at angle close to $\phi_r/2$, with residual value of ϕ_r (section 26).
London Clay has $\phi_r = 20°$; slopes are stable at $< 10°$, and do not exist at $> 12°$.
Old slides are at residual strength; reactivated mainly by head loading or toe removal.
Failure surface in homogeneous soil is a slip circle.
Critical slip circle, of lowest safety factor, is found by tedious calculation of all possible circles, summing data on slices and using iterative methods; now always by computer.
Back analysis obtains soil strength parameters by stability analysis of failed slopes when safety factor = 1.

PROGRESSIVE FAILURES

Clays are brittle and lose strength as they fail.
Brittleness = % loss of shear strength, from peak to residual.
Clay soil in slopes too steep or too high is locally overstressed; deforms and loses strength, passes load to adjacent soil; shear zones grow and coalesce; overall strength declines to ultimate failure of slope.
Scale of progressive movement and partial failure depends on brittleness.
At intermediate stages, some of the soil is loaded at peak strength, some is losing its strength in post-peak deformation.
Mean strength during failure is close to ϕ for peak strength with cohesion close to 0.
Progressive failure may take years. Many railway cuttings in London Clay have failed after 50–100 years, before reaching residual strength.

Coastal landslide on the Isle of Wight; a soft clay has failed between vertical beds of sand, developing into a mudflow with typical arcuate pressure ridges.

FLOWSLIDES

Soil, clay or rock debris may fail as flowslide where material behaves as a liquid; water content is above liquid limit. Usually due to decrease in strength, not to increase in water.
Liquefaction: total loss of strength due to undrained restructuring. Disturbance, by shearing or vibration, destroys soil skeleton; with loss of grain contract and decrease of porosity, soil load is transferred to pore water; water pressure > normal stress, so effective stress = 0, and soil acts as a liquid. Drainage reduces pore water pressure, allows grain contact and thixotropic recovery of strength.

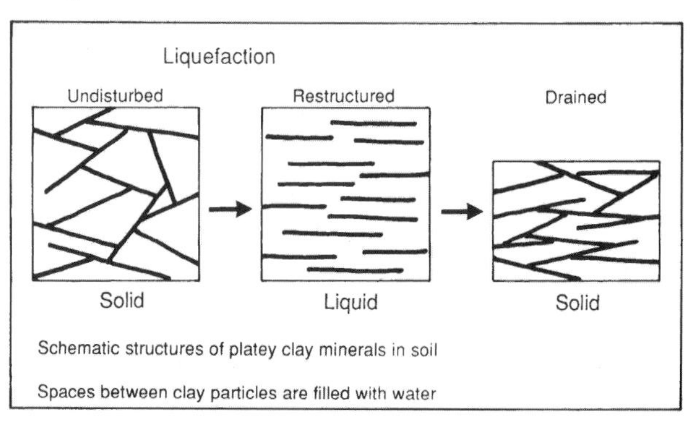

Schematic structures of platey clay minerals in soil
Spaces between clay particles are filled with water

Fluidization: develops in moving materials, notably rockfalls and pyroclastic flows. Grains continually bounce off each other with no permanent grain contact or strength. Pore fluid may be water or air. May also involve hovercraft effect on cushion of trapped air. Stops when movement reduces to critical point.
Flowslides are extremely mobile; move over low gradients. Most develop from smaller initial slides, notably slope failures in sensitive clays, slides in non-cohesive tip debris, rockfalls in mountain regions. Also due to earthquake vibration.
Fluidized rockfalls include Frank and Yungay (section 32) and Saidmarreh; also many large slides which block deep valleys in Himalayas. Most violent are known as sturzstroms.

SAIDMARREH ROCKSLIDE, IRAN

Prehistoric slide, perhaps largest in world.
Limestone slab, 20 000 M m^3, slipped on dip of 20°.
Fluidized; mobile enough to travel 16 km across valley floor, at > 300 km/h, and up over ridge 450 m high.
Slide debris first thought to be glacial till.

SOLIFLUCTION

Downslope movement of saturated debris.
Periglacial conditions (section 16) in Pleistocene caused numerous slope failures in Britain.
Solifluction of active layer widespread on slopes > 4°, notably in clays, mudstones and chalk.
Postglacial melt of permafrost permitted drainage and marginal stabilization, leaving shear surfaces in the head with residual strength of $\phi_r = 8–15°$.
Many slides reactivated since recent deforestation.
Sevenoaks bypass, Kent, had to be relocated across soliflucted slope; Pleistocene flows on slopes of 2–7° consisted of head few metres thick overlying clay; upslope cutting sides were impractical to stabilize.
Any slope > 5° in clay, which was in Pleistocene periglacial zone, is likely to have head debris prone to reactivation.

QUICK CLAY SLIDES

Sensitivity, or extreme undrained brittleness, is loss of strength when sheared.
Sensitivity = ratio of compressive strengths, undisturbed : disturbed.
Sensitive clays have sensitivity > 4.
Quick clays have sensitivity > 16.
Highest sensitivity in clays or silty clays, due to realignment of clay plates (mainly illite) and silt grains.
Fine grain and low permeability hinder drainage and allow liquefaction to develop fully.
Leda Clay of Eastern Canada is the classic example of quick clay, along with similar clays from Norway and southern Sweden.
Marine clays formed around margins of Pleistocene ice sheets; then uplifted by glacial unloading; now form low valley-floor terraces with steep edges where cut into by postglacial rivers.
Clay particles were originally flocculated into clumps due to bonding in saline porewater.
Leaching by modern rainfall removes salt and interparticle bonding, leaving metastable structure.
Liquefaction is caused by small slope movements in terrace edges when porewater has < 1 ppm salt.
Flowslides develop and rapidly expand headward.
Stabilization may be possible by brine injection.

NICOLET FLOWSIDE, CANADA, 1955

Typical failure of terrace in sensitive Leda Clay.
Rapid headward growth carried away buildings, and left large arcuate scar 10 m deep.
Flowslide debris spread 250 m into river.

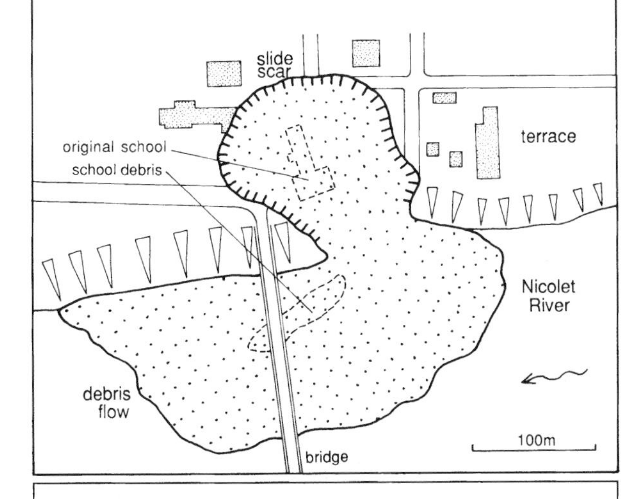

TURNAGAIN HEIGHTS SLIDE, ALASKA

Clay terrace 25 m high with 8 m gravel cap.
Clay has 40–53% < 2 μ diameter; sensitivity 10–40.
Major earthquake in 1964 had local intensity VIII, with strong motion for > 4 minutes; unusually long duration.
After 90 seconds of vibration, clay liquefied.
Major translation slide; movement in lower 8 m of clay.
Slide area extended to 50 ha; 75 houses wrecked, some moved 150 m.
Ground stable again after earthquake vibration ended.
Clay sensitivity known from laboratory tests in 1959.
The hazard had been recognized, but had been ignored as it would only materialize in an earthquake; but this was almost inevitable in Alaska.

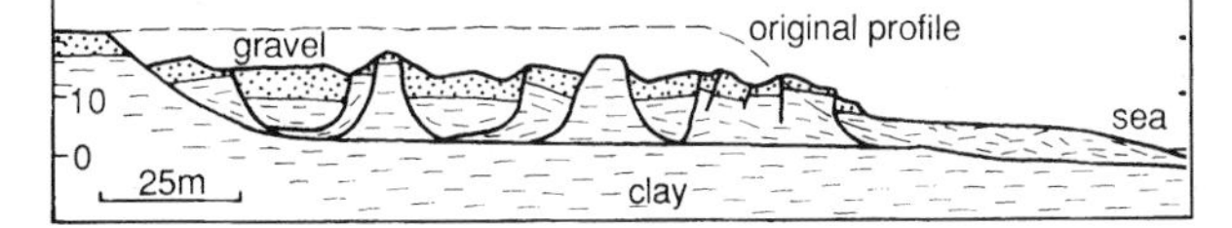

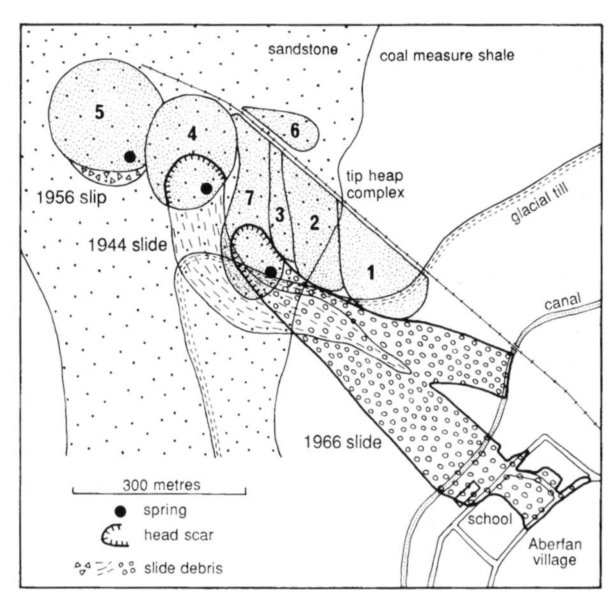

ABERFAN TIP FAILURE, WALES, 1966

Best known of many similar colliery tip failures because it struck the village school and killed 112 children.
Failed material was artificial but causes lay partly within geology of natural ground.
Multiple event with rotational slip followed by flowslide and mudflow.
Location of tip 7 (which failed) was very unsuitable.
Tip overlay natural spring from sandstone.
Springline along base of Brithdir Sandstone well known as site of many previous tip failures in Welsh Valleys.
Tip extended over sheared debris of earlier slide from higher tip.
Rotational slip prompted by head loading of slope – common event on tips built from top.
Large scale of slip due to saturation of debris by spring beneath; three previous failures were over springs, no failures of dry tips.
Slow movement; 6 m headscar developed in a few hours before formation of the flowslide.
Mining subsidence ground strain produced local extension zone, increased rock fracture permeability and spring flow, raising water pressure within tip.
Flowslide formed in saturated debris restructured in rotational slip, with reduction of porosity and strength decline to residual.
Liquefaction as debris could not drain; low permeability due to high content of fines from mine washing plant.
Flow of 110 000 m^3 debris, moved 610 m on 13° slope.
Rapid consolidation when the flow stopped.
Mudflow formed in some of debris with water released from sandstone when glacial till stripped by main slide.
Cause: essentially the saturation of the undrained tip debris, which was placed over a spring – due to a total lack of ground investigation prior to tipping.
Other factors increased the scale of the disaster.

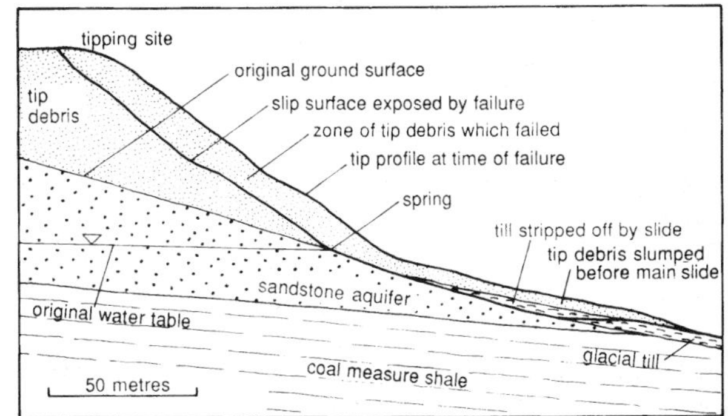

35 Landslide Hazards

POTENTIAL SLOPE FAILURES

Most failures of rock slopes are related to planar weaknesses with unfavourable orientations.
Wedge failures most likely where line of intersection of two fracture planes dips < ϕ and daylights in slope. Best interpreted graphically by stereoprojection.
Hazard zones. A failure potential can be recognized, but incomplete data on buried rock structure means that stability analysis can only be estimated, and risk assessment is subjective dependent on strengths and structures assumed.
Time of failure cannot normally be predicted except that it will probably be during or just after a rainstorm. Size and speed of future event is rarely predictable.
Small unsafe slopes may be economically stabilized.
Large unstable slopes are best avoided if possible.

LANDSLIDE HAZARD MAPPING

Can be effective for route planning and land use zoning.
Assessment factors include:

- Rock type, structure and strength;
- Soil type and plasticity;
- Slope angle and shape;
- Drainage state and level of water table;
- History of previous landslides;
- Land use, including vegetation type and change.

Local data is the basis for any hazard evaluation, notably the slope angles known to fail in each rock and soil type, and any particular rock structures.
Key factor is recognition of old, inactive slides.

OLD LANDSLIDES

Low stability because shear surface has reduced to residual strength with little or no cohesion. Reactivated with no peak strength to overcome.
Stable slopes therefore close to $\phi_r/2$ in saturated soils and debris; slightly steeper in rocks and where water table cannot rise to surface.
Recognize old slides by surface features:

- Irregular hummocky ground on slipped material;
- Lobate plan of debris and solifluction flows;
- Unsorted slide debris which may resemble glacial till;
- Concave upper slope and/or convex lower slope;
- Head scars, rounded in soils, angular in rock;
- Back dips in slipped blocks due to rotation.

Shear zones: in rock, may have parallel shear fragments; in soils, mostly remoulded, soft to medium stiff, wet, silty clay, 1–50 mm thick.
Slip surfaces, in rock and soil, may be clean breaks with or without polished surfaces and/or slickensides.
Identification of slip surfaces may be difficult in boreholes due to core loss or confusion with simple fractures.
Investigation of shallow slides is best with trenches or trial pits; trench wall cut smooth with knife normally shows slip planes in clays after they have opened due to shrinkage on drying; important in recognition of solifluncted head.

ACTIVE SLOPE MOVEMENT

Surface signs include: fresh scarps or terraces; new ponds, undrained hollows or pressure ridges; fresh sharp rock fractures; distorted tree growth.
Locate depth to slip surface by offset of borehole which is unlined or cased with flexible plastic, or by acoustic emission profiling.

SLOPE HEIGHT

Cohesion allows slopes to stand at angles > ϕ (angle of internal friction of the rock or soil).
Frictional resistance, effective stress and driving force are all functions of mass, increasing with slope height; cohesion is a function of area, independent of height, so it has less proportional effect in high slopes.
Low slopes can therefore stand at steeper angles than high slopes in same material.
Intact rock (horizontally bedded) will stand in vertical cliffs of height limited only by UCS; coastal cliffs in weak chalk stand 150 m high.
Height and slope of natural and cut faces is limited by fractures – mainly their orientation, also density, roughness and shear resistance.

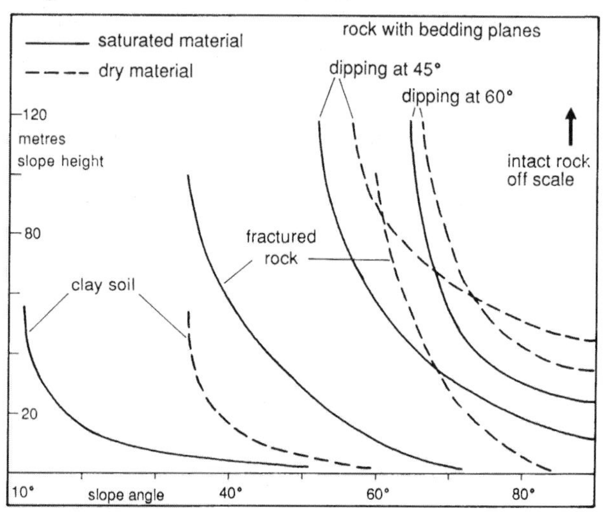

Influence of slope height on stable slope angle

High slopes in dry materials all have stable angles which approach the lesser of ϕ or the dip of any structural weakness.
All stable slope angles are reduced by saturation (roughly halving the tangent of the angle); saturated values apply to long-term stability.
Shallow surface slides and isolated rockfalls are independent of slope height.

Karakoram Highway crosses over 500 km of mountains between Pakistan and China, with steep slopes in geologically active terrain; some part of the road is closed on average one day in two by rockfalls, debris flows and mudslides landing on road; most blockages are cleared within a few days, and repeat failures are inevitable; stabilization of high slopes above the road is impractical and uneconomic; roadbed is well engineered so it rarely fails except on a very small scale.

MAM TOR LANDSLIDE

Glacially oversteepened slope on shale and sandstone in English Pennines.
Slide 300 m wide, 1000 m long, > 3000 years old.
Upper part is multiple rotational landslide of bedrock slices, with road stepped over minor head scarps.
Lower part is debris flow, creating wavy road profile.
Road across the slide has been closed since 1977.

Stability analysis considers slices extending down to slide, each 1 m wide, each broken into sections of uniform properties.

Factor of safety = FS = shear resistance/driving force

$$FS = \frac{\Sigma c'l + \Sigma (W\cos\alpha - ul)\tan\phi'_r}{\Sigma W\sin\alpha / (1 + d/b)}$$

- c' = apparent cohesion; taken as 0 on reactivated slip surfaces
- l = length of slip surface in vertical sided section
- W = mass of slice; unit weight = 20 kN/m^3
- α = inclination of slip surface
- u = pore water pressure = height of water table above slip surface
- ϕ'_r = angle of frictional shear resistance; taken as residual value of 14°, from tests and back analysis of nearby slides
- d/b = slide depth/slide width; for edge drag due to lateral earth pressure.

Calculated values of FS for whole slide are close to 1·0; these are correct as slide is in critical state, just moving; so assumed values of c and ϕ are good.

- Upper slide: analysis of typical slice section: where d = 20, u = 10, α = 13°; then FS = 0·86
  if water table is lowered, u = 5; then FS = 1·01
- Lower slide: analysis of typical slice section: where d = 15, u = 12, α = 7·5°; then FS = 1·19
  if water table rises to surface, u = 15; then FS = 0·99
- Whole slide: FS reduces by 0·05 for every 1 m rise of water table.

Movements of around 0.7 m occur one winter in every four, when threshold groundwater levels are exceeded – by a winter month with rainfall 60% over average following a year of above-average rainfall.
Drainage with deep sub-horizontal boreholes or adits (see section 36) would stabilize the slide effectively.
Earth shift from head to toe would have minimal effect as debris could flow round any toe weight (or anchor).

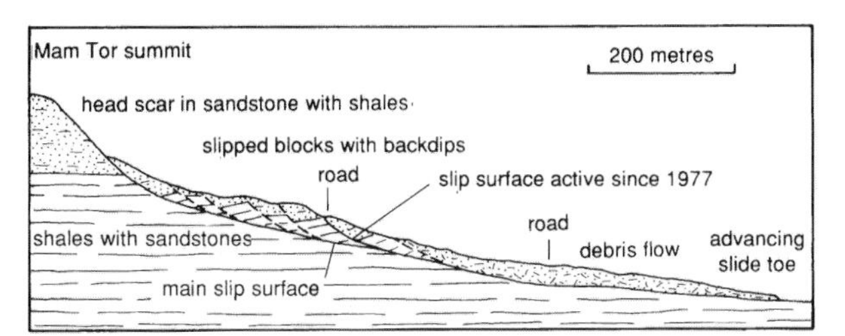

STONE FALL

Almost all rock faces, natural and cut, produce random falls of small rocks due to natural weathering.
Periodic face cleaning by roped-access technicians can reduced hazard; safety catches still needed where roads and buildings are threatened; these are generally cheaper than face stabilization works.
Rock trap ditches at foot of slope: bank 1·5 m high, best topped by fence or dense bushes, in front of ditch 3 m wide with earth floor for impact absorption, good for most faces < 20 m high; 5 m wide ditch for higher faces, and higher bank for slopes around 60°.
Smaller ditches or fences alone below low cut faces.
Gabion walls on low gradients are very effective and economical traps.
Rock catch nets on steep slopes: wire or rope netting hangs in catenary sag between hinged supports, with cable anchors to rockbolts using cable brakes (loops designed to slip through clamps); designed to deform to absorb impact energy.
Rockfall shelters protect roads beneath very loose slopes; massive concrete with roof covered by blanket of crushed rock (or sloping like avalanche shelters).

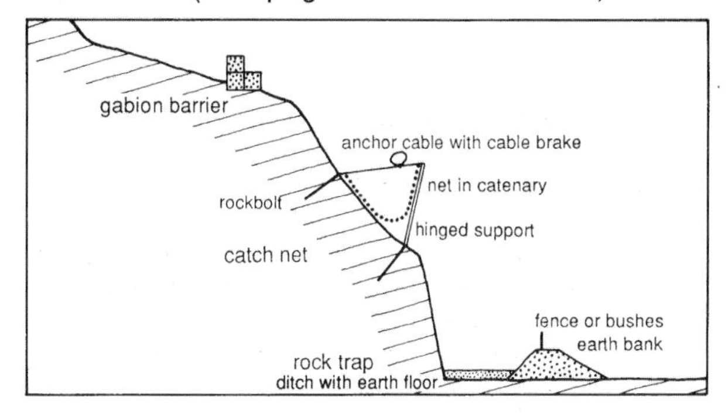

LANDSLIDE MONITORING

Various means of instrumentation; wide range of cost.

- Surface surveying between fixed reference points.
- Crack dilation measured between studs glued to rock.
- Standpipe or electric piezometers, manual or automatic reading of water table.
- Boreholes with installed inclinometers.
- Horizontal or vertical borehole extensometers.
- Photoelastic or electric load cells on installed anchors.
- Geophones record acoustic emissions (ground vibration due to movement).

Simple, rugged instruments are most reliable.
Reading and interpretation must be over a long time.
Monitoring shows any acceleration of movement; interpretation is difficult as critical values are unknown unless there is documented history of previous events (as in some large quarries). Vaiont slide (section 33) was monitored, and failed unexpectedly.
Extensometers, geophones and electric trip wires can be linked to automatic warning systems, as in some railway cuttings in Sweden.

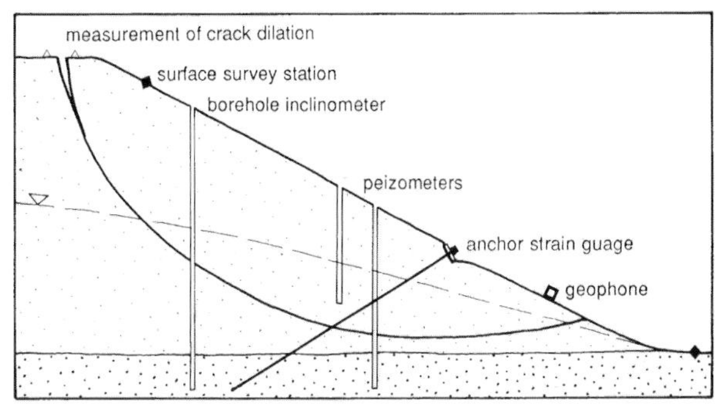

36 Slope Stabilization

Slopes may be stabilized by one or more of:

- Modifying the slope profile, where possible;
- Supporting or anchoring the existing profile;
- Improving or draining the slope material.

PROFILE MODIFICATION

Added material below neutral line, also removal from above, increase stability. Neutral line is where slip surface is horizontal below undrained slide; displaced to where dip of slip surface = ϕ in drained state.

Berm ledges, about 5 m wide on steps 10 m high, redistribute load and stabilize circular slips in weak rock. Small failures on steep faces of steps land on berms harmlessly.

Hanging blocks, slabs and wedges, which rest on daylighting, unfavourably orientated fractures, can be removed; may leave asymmetrical road cuttings in beds with dip > ϕ.

Toe weight is effective, especially where lower end of slip surface turns up. May be mass concrete, rock fill, earth bank reinforced with geogrids, or thick crib walls. Size must restrain formation of new underriding slip surface reaching beneath new toe.

Unloading head of slide generally has less effect.

Original cause of failure should be removed where possible: river bank erosion control or sea wall to prevent toe removal.

DRAINAGE OF SLIDES

Pore water pressure is critical to slide stability, so drainage is usually very effective, and is the only economical method to stabilize large slides in natural slopes.

Surface drains: concrete diversion ditches intercept surface flows; drains on slides reduce infiltration.

Shallow drains: stone drains in trenches 1–2 m deep lined with geotextile; have limited effect by reducing soil water; deeper counterfort drains also provide shear resistance.

Deep drains: most effective; mined adits with leaky walls and sealed floor, or boreholes with perforated casings, inclined to drain out from slide toe.

Relief wells: drain up or down through aquiclude; need pumping unless drain into lower aquifer. Some London Clay slopes have been drained into lower sand, through 100 mm boreholes filled with sand in polypropylene tubing, on 2–5 m centres.

Interception tunnels: cut behind slide to reduce groundwater inflow; used in 1800 to stabilize failing slopes above city of Bath.

Impermeable clays respond poorly to normal drainage. Electro-osmosis or heating with ducted hot air can improve clay stability, but are expensive.

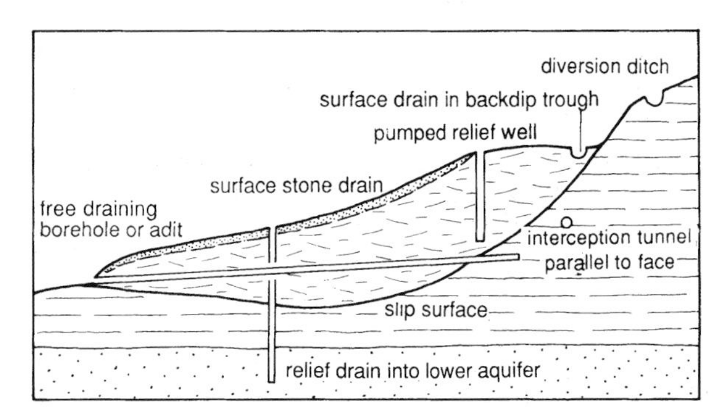

LLANDULAS LANDSLIDE

New road along North Wales coast and new marine defence works were sited to act as toe weight, improving stability of old slide and restraining formation of new underriding slip surface.

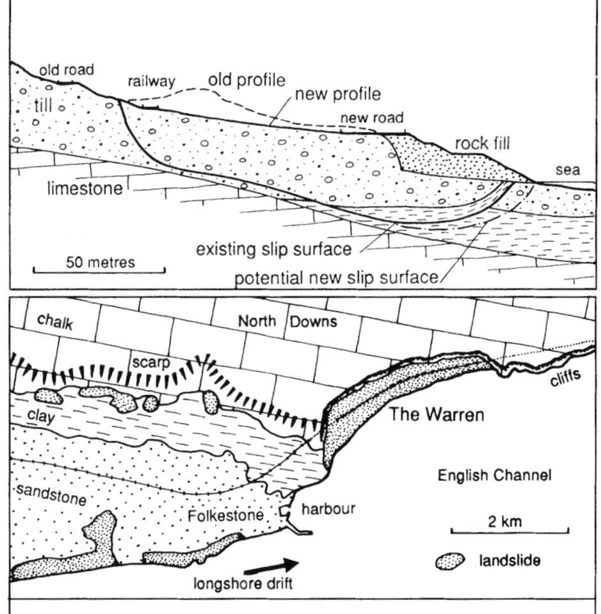

FOLKESTONE WARREN LANDSLIDE, KENT

Multiple rotational coastal landslips in chalk over clay, crossed by mainline railway.

- High winter water tables and relaxation of over-consolidated clay cause renewed movements; largest slips are triggered by falls from chalk cliffs onto slide head.
- Toe erosion increased due to beach starvation after harbour wall, extended in 1905, traps longshore drift.

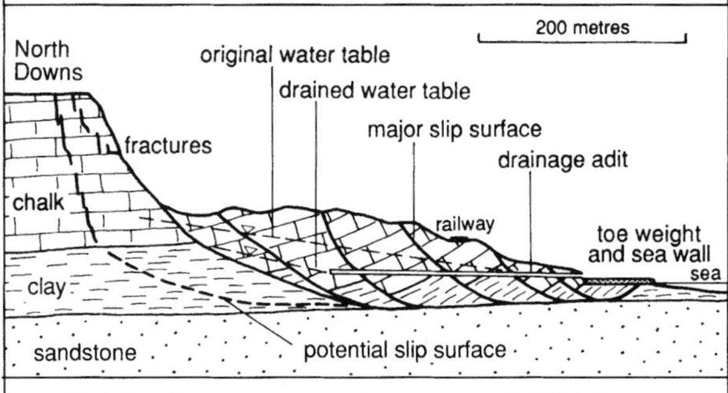

- Stabilization measures placed since 1915 failure, and added to since: mass concrete toe weights also act as sea wall to reduce toe erosion; drainage adits cut in from sea wall lower water table.
- Movements are now reduced to very low rates.

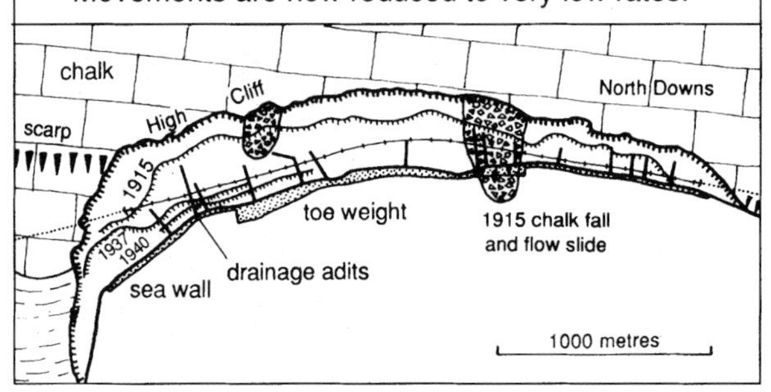

MASS SUPPORT

Retaining walls: common practice and successful on small slopes and cut faces, but not on larger slopes.
Large unstable slopes are not easily retained; it is difficult and usually uneconomic to build and found massive retaining walls at the toe of major natural landslides or large unstable slopes.
Concrete walls: need sound footing; prevent from rotation by buttresses, rock anchors from near top and/or base, or deep foundations; weep holes to permit drainage; masonry facing improves appearance.
Dental masonry: hack out weak zone of rock and refill with cemented masonry or concrete with stone facing; may dowel into rock.
Gabion walls: cheap to install, can retain soil slopes and act as toe weight.
Sprayed concrete (shotcrete): may be used with rockbolts (as in tunnel support, section 38); spray over bolted reinforcing mesh, or use fibrecrete containing 50mm lengths of steel wire within the sprayed mix to give tensile strength.

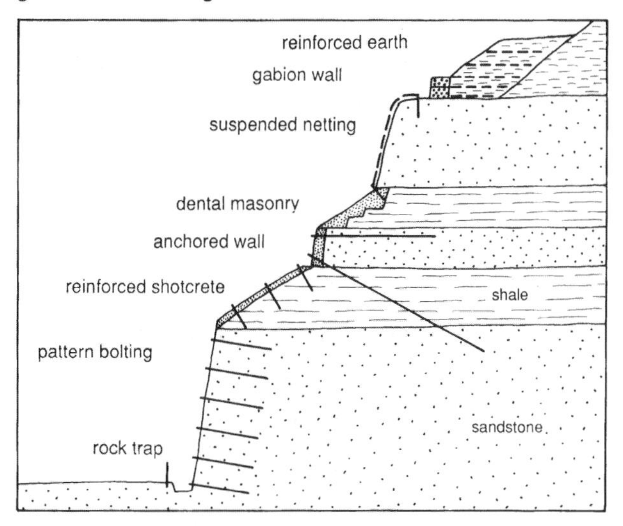

GROUND IMPROVEMENT

Vegetation cover reduces rainwater infiltration, and root mat provides tensile strength in soils – but is destructive on rock faces, as roots force open joints.
Geotextiles, geogrids or wire mesh, anchored to bolts, fixes surface and catches small stonefall; plants grow through and provide long-term strength; biodegradable jute mesh provides short-term support.
Weathering protection by sprayed concrete or fibrecrete, or by chunam (spread mortar); any cover must have drain holes.
Pattern bolting creates thick stable layer by tightening the natural rock joints (section 38), and also reduces water infiltration.
Grouting fractured rock is expensive and rarely applied; may use to stabilize scree.
Liming clays reduces plasticity; improve unstable sodium montmorillonite by change to calcium variety.

The Folkestone Warren landslide.

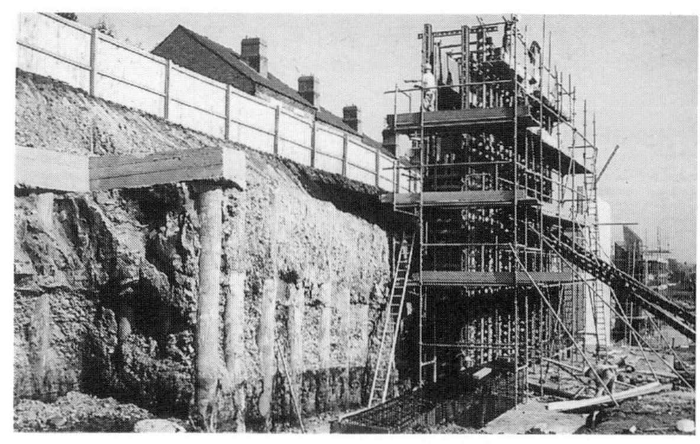

Retaining wall under construction with frontal toe and shear key beneath for new road cut in Derbyshire. Exposed bored piles provided temporary support.

GROUND ANCHORS

Tensile support may be provided to oppose directly the landslide driving force.
Rockbolts: steel bars, 25 mm diameter, 3–10 m long, in drilled holes, fixed at inner end by resin or expansion shell, 100 kN load, tensioned to 60 kN. Main purpose to increase normal stress across joints; cannot be placed in heavily shattered rock; isolated spot bolts to retain individual blocks of rock.
Grouted dowels: steel bars in drilled holes through joints provide direct shear resistance.
Bored piles: act as large dowels, but limited success. Concrete piles 6 m long, 1·2 m diameter, placed across slip surface of Portuguese Bend landslide, Los Angeles, had no measured effect; some rotated, some sheared, slide flowed round others.
Rock anchors: multiple steel cables, 10–40 m long, in ribbed protective casing for installation in drilled 100 mm holes, up to 2000 kN load, tensioned to 60%; fixed length about 5 m resin bonded into rock. Provide tensile support and tighten rock fractures. Most effective where installed at angle of f above slip plane.
Anchored walls, flexible or rigid, distribute load from anchor caps onto weak landslip material.

HOAR EDGE CUTTING

Shallow slide initiated during construction of Pennine motorway, stabilized by lowered profile and rock buttress. Subsequent movement required anchored wall – crushed rock in geotextile faced with shotcrete, with 41 cable anchors, each 1000 kN, spaced 1–3 m, bonded for 6 m into stable sandstone.

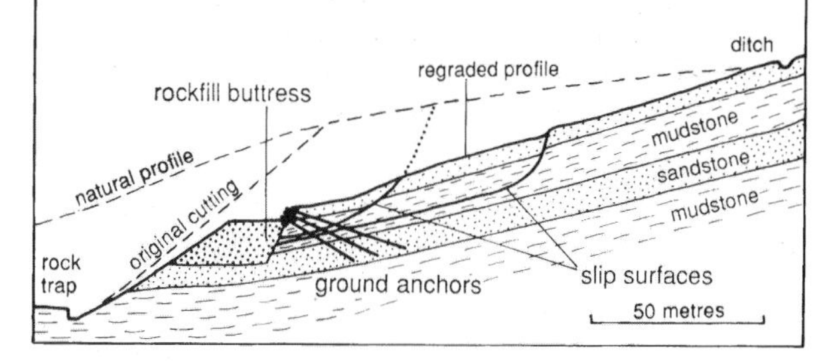

SHEAR KEYS

May be used to stabilize toes of slides in weak soils. Trenches filled with granular material (with high ϕ) reach through slip surface into stable ground.

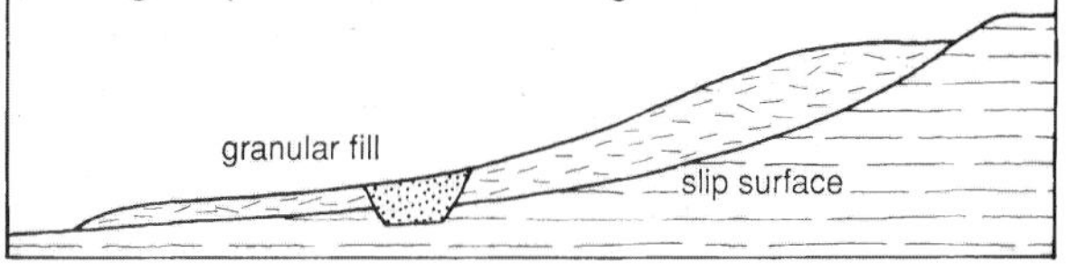

37 Understanding Ground Conditions

The outcome of a successful ground investigation is a broad understanding of the geological conditions of a site, and the implications that these may have on any planned engineering activity.
This requires an overview which may rely on geological experience – not a problem on large projects employing specialist engineering geologists, but maybe critical on small projects with a limited geotechnical team.

Ground conditions are only fully understood when different parameters are integrated into a concept:

- Nature and strength of the rocks and soils, and the difficult ground that any of these may provide.
- Fracture conditions of the rocks, with respect to the rock mass strengths that they determine.
- Geological history, and stress conditions in the ground, critical to underground engineering.
- Groundwater conditions, and slope stability with respect to pore and joint water pressures.
- Quaternary evolution, notably rock deterioration by weathering, and de-stressing by erosion.
- Man-made impacts on the ground, including any contamination of brownfield sites.

TOTAL GEOLOGICAL MODEL
This concept has been introduced to present the broad picture of ground conditions. It develops an overview, that is the normal outcome of a geologist's thinking, but is rarely foremost in an engineer's perception.
3-D drawings of ground models incorporate all the individual components of the ground conditions.
The total model matrix has three types of models:

- tectonic – that outline the background data;
- geological – to provide the broad ground picture;
- geomorphological – with the near-surface details.

Good model drawings demand some artistic ability, but even rough sketches expose deficiencies in the data on ground conditions of a site, and they focus attention on potential engineering problems.
For a small site, a model may be only a thumbnail sketch, but multiple detailed models are becoming increasingly important on large projects such as highways, pipelines and new town developments.

Outcomes of an engineering geology investigation:

- One or more conceptual ground models – which are very helpful to project managers and engineering personnel who do not have a full geological or geotechnical background.
- Identification of areas of difficult ground, and the scale of their potential geohazards.
- An engineering geology report in two parts –
  Part 1 – factual data: with all geological records;
  Part 2 – interpretation: of the ground properties and conditions related to the construction project, potential problems, and the limitations of the data.

Too often, inadequate or misdirected investigations:

- rely on boreholes and trial pits that expose only a tiny fraction of the ground under a site;
- examine parameters not relevant to the problems;
- fail to discover critical ground conditions.

'Unforeseen ground conditions' are, in most cases, only unforeseen because nobody had looked for them.

RECOGNITION OF GEOHAZARDS
The steps in good geotechnical risk management:
Create a team of geotechnical experts.
Gather all available data on ground conditions.
Establish the likely/possible forms of construction.
Identify the potential geotechnical hazards.
List and rank risks for each part of the project.
Relate risks to engineering options and to costs.

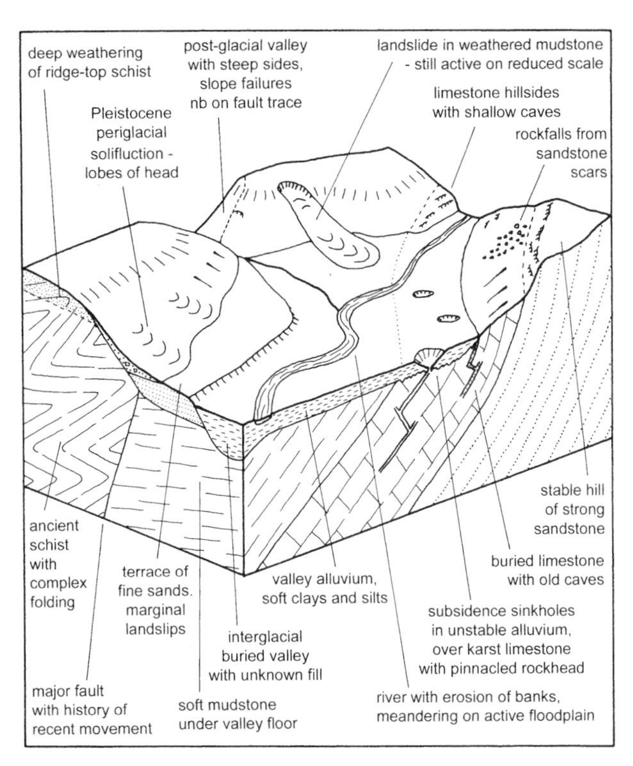

A total geological model for an upland site that was influenced by variations in the Quaternary climate.

The models in a ground investigation: they are best created after the initial stage; this should allow a more efficient main investigation with boreholes and testing; and they should evolve as investigations and site works progress – in the style of an observational approach.
The models should provoke both directed and lateral thinking, and so lead to a more balanced and effective ground investigation.
Even if the models are not pursued in every detail, their concept is applicable to every engineer who has to achieve an adequate understanding of the ground conditions at his project location.

Progress in Engineering Geology
This is well demonstrated by the Glossop Lectures for the Engineering Group of the Geological Society.
Fookes, 1997, introduced the concept of a broad understanding through the Total Geological Model.
Hoek, 1999, moved towards a more numerical approach, particularly suited to underground works.
Hutchinson, 2001, brought geomorphology into the forefront of understanding ground conditions.
Successful engineering geology should incorporate all three approaches, as appropriate to each project.

Geomorphology is the study of the ground surface, notably the Quaternary evolution of the landscape.
The Quaternary is only a tiny fraction of geological time, but it was so recent that its processes of erosion and weathering have a disproportionate impact on the state of the ground today.
Geomorphology is often overlooked by engineers, but simple surface observations can provide data invaluable to the interpretation of sub-surface ground conditions. Primary geomorphological mapping plays a key role in highway engineering in the difficult terraines of undeveloped countries.

Quaternary processes accounted for a sequence of changes within the ground conditions by mechanisms that may not be active at present –
Weathering and degradation of near-surface soils and rocks, notably related to Pleistocene periglacial zones.
Drift deposits with respect to Pleistocene glaciation limits, temporary lakes and river drainage patterns.
Erosional removal of cover rocks, causing stress-relief opening of fractures in rock, notably on hillsides due to valley incision (and also overconsolidation of clays).
Slope instability in terms of past drainage states and the contrasts in slope processes in changing climates.

Brownfield Sites

Increasing demand for building land, and a shortage of greenfield sites, creates a need to re-use 'brownfield' sites – derelict land, or 'made ground' that includes old opencast mines, backfilled quarries, old industrial sites and disused waste dumps.
Of these sites, about 65% are contaminated with toxic metals, chemicals, organics and/or hydrocarbons.
On clean made ground, settlement is main problem.

SITE INVESTIGATION

On brownfield sites, this is more than a normal ground investigation, as many legal, historical and environmental factors have to be considered; it is a specialist field, where the concept of a total geological model is particularly appropriate.
Staged investigation is best on an unknown site; with pits and trenches to sample solids, hollow probes to test gases, and boreholes to monitor leachate flow.

CONTAMINATED LAND

This includes any site where buried substances may become accessible and so present a health hazard.
Each site is different, and may respond differently to disturbance, notably by migration of leachates or gas; remediation is only needed where the risk is unacceptable, but limits are not easily defined.
Harmful materials may have to be removed to a safe site, may be buried on site (except oils) under clean soil cover, or may be isolated by grout cut-off and deep burial. Total clean-up may be cost-prohibitive.
Organics may be reduced by on-site bioremediation.

SETTLEMENT OF FILL

Uncontrolled fill may have high potential compaction. For loading of 100 kPa (house strip footings), Young's modulus E varies from > 10 MPa for dense rockfill to < 1 MPa for domestic waste. Creep can last for years.
Easy field test of settlement is a sand-filled skip left on site for a month; most movement is very rapid.
Normal to use raft foundations for houses on soft fill.
Main hazard is long-term differential settlement (tilt) over variable fill. Buried opencast highwalls and quarry faces must be traced and avoided; tilt could be excessive and could break a raft.
Inundation collapse is loss of volume when fill is first saturated, by changed drainage or rising water table after mine-pumping stops. Loss of thickness may be 1% on compacted rock debris, over 7% on non-engineered mine waste, and higher on some refuse.

TREATMENT OF MADE GROUND

Various methods of ground improvement (section 27) can reduce long-term settlements.
Preloading effectively compacts the ground to a depth that is about 1.25 times the depth of surcharge.
Dynamic consolidation is effective to depths of 9 m in sand and rockfill, or 6 m in clay or mixed refuse.
Vibroreplacement stone columns can improve any fill.
Pre-inundation may treat dry fill prone to collapse.
Methane, derived from buried domestic waste or from coal-bearing rocks, should be drained to the air, or may be tapped and burned for power production.

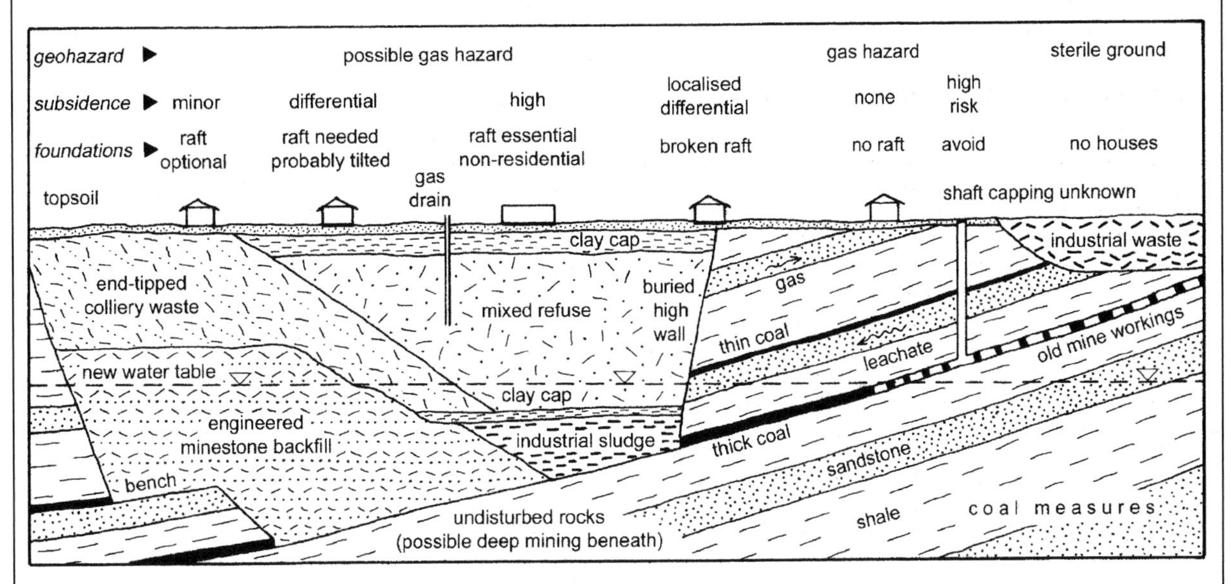

Varying conditions on brownfield land with past mining and landfill, and the state of houses built on rafts.

38 Rock Excavation

EXCAVATION METHODS

Method relates to rock strength and fracture density.

- **Direct excavation**: possible in fractured rock of mass class V (section 25) and in all soils; using face shovel, backhoe, clam shell grab or dragline.
- **Ripping**: needed to break up slightly stronger rock, roughly class IV; using tractor-mounted ripper, or breaking with boom-mounted hydraulic pick (pecker).
- **Blasting**: generally required in stronger, less fractured rock. Class III rock is loosened in the ground by undercharged blasting in some quarries; on urban sites can be broken by hand-held pneumatic drill or by pecker. Massive rock of moderate or high strength, class I or II, needs to be fractured normally by blasting; where blasting is unacceptable, breaking by pecker or hydraulic breaker is very slow.

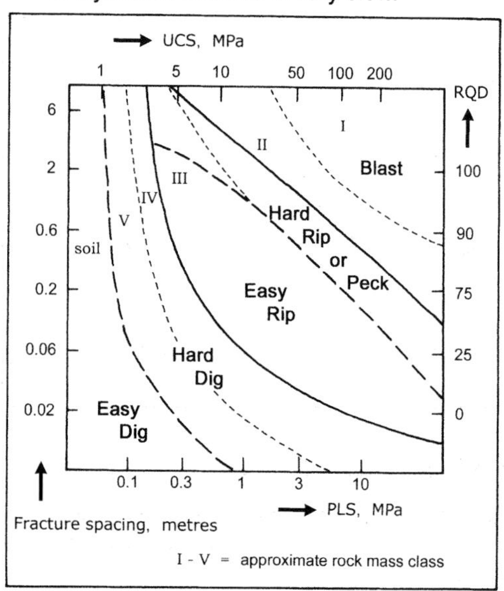

CUT SLOPES IN ROCK

Sound rock can be cut to vertical faces; normally raked back by 10° and benched at 10 m intervals to improve safety. Higher benches cannot be drilled accurately; berms act as rock traps, on highway cuttings and in working quarries.

Inclined fractures are main hazard, notably dipping 30–70° and daylighting in face. Dips > 50° normally require cutting face back to clean bedding or fracture.

Shale beds may weather and undercut slopes in strong sandstone or limestone.

Slake durability test measures % retained intact through 10 minutes of wet/dry cycles in standard drum apparatus. Most rocks have values > 90. Values < 50 for shales indicate susceptibility to weathering and long-term slope degradation.

Hillside excavations may undercut unstable weathered rock, old landslides or soliflucted head.

Floor heave is rare in rock excavations; only likely if unloading stress > 6 x shear strength.

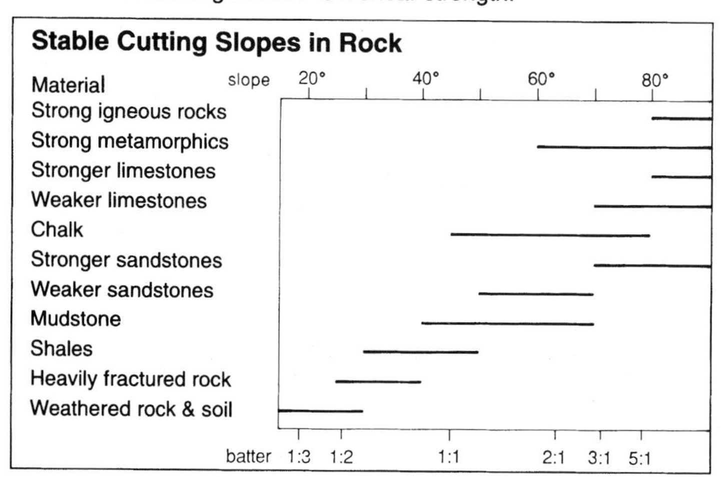

Guidance Values for Long Term Stable Slopes

Values are reduced for slopes with poor drainage, steep dips or structural loading, in areas of deep weathering, mining subsidence or seismic activity, and for unbroken heights > 20 m.

PANAMA CANAL, INSTABILITY OF CULEBRA CUT

Deep cut planned as vertical, completed with 15° sides. Repeated slides in saturated shales and tuffs only overcome by American use of steam shovels.
Excavation: planned 46M m^3; completed 170M m^3.

CUT SLOPES IN CLAY

Drainage changes stability over time where face is cut into clay with initial water table near the surface.

1. Excavation permits stress relief, pwp decreases.
2. Pwp rises to regain equilibrium (drained state); strength and stability therefore decrease.
3. Slope ultimately drains (or is artificially drained) to new lower water table;
   reduced pwp then increases stability.

Rate of change depends on permeability.

Temporary faces in clay can stand vertical to height = 4 x cohesion/unit weight, less the depth of any tension fissures. Walls shear at critical values; base failure undercuts propped faces at slightly greater heights.

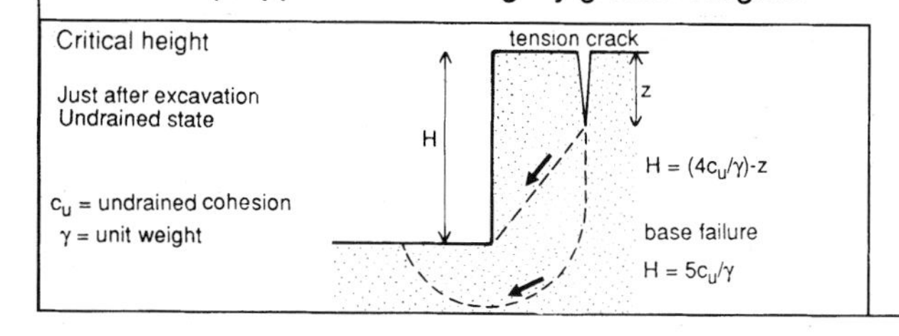

Premature failure occurs where stability is due to temporary pore water suction; failure may be in minutes or hours so faces are battered back for longer safety.

London Clay, unweathered, is cut to 65° slopes to 8 m high where small slips can be tolerated; reduce to 20° for unsupported faces below buildings.

Stiff glacial till may stand close to vertical for some months at less than critical height, so retaining walls can be built in front. Weep horizons on sand layers cause instability.

Lateral stress relief in slopes cut in over-consolidated clay may cause outward movement.

Settlement adjacent to stable cut slope may be 1–2% of excavation depth, reaching 2–4 x depth back from crest of slope.

material	cohesion	critical height, H	
		unfissured	fissured
Soft clay	25 kPa	5 m	3 m
Firm clay	50 kPa	10 m	6 m
Stiff tile	12 kPa	24 m	15 m
values for typical fissure depth = z = 1·5 c/γ			

OPEN FACE BLASTING

Drilling: method should relate to rock toughness – highest in fine grained igneous rocks.
In soft rock use pure rotary drills.
In hard rock use rotary percussion drills with tricone or roller bits, or down-hole hammer.

Drill holes: 50–100 mm diameter (D), preferably at 10–15° from vertical.
Burden is distance of drillhole line from free face, therefore thickness of rock to be moved. Ideally should be 30–40 x D, generally 2–4 m.
Spacing along line = 50 x D, generally 3–5 m.
Hole depth = 2–4 x burden, commonly 10–15 m.
Subdrilling needed below grade to depth = burden/3.

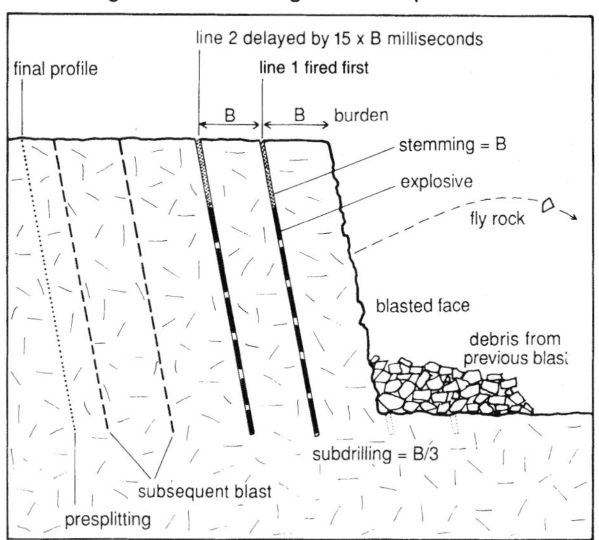

Charging: generally with 0·4–0·6 kg ANFO per m^3 rock.
Stemming is sand filling of hole above explosive, to depth = burden; to reduce fly rock.
Firing is usually by electric detonator caps; electric delay caps allow firing of multiple lines in one round, each with optimum burden; so second, inner, line is delayed by 5 milliseconds/m of burden.
Fragmentation of rock is best determined by field trials; improve by smaller spacing and burden and short delays which superimpose vibrations.
Blast energy travels along joints; quarry faces are most economical parallel to joints.

Controlled blasting of final perimeter line should leave clean wall free of blasting fractures.
Presplitting: holes with spacing only 10–20 x D; charged with 10% normal, low density explosive, decoupled (not rammed tight against drillhole walls, so reducing fracturing); fired simultaneously with high burden (before removal of main bulk), to create a single clean break linking the holes.
Line drilling: holes with spacing only 2 x D, not charged, so main blasting can work back just to line of perforations; expensive, used only on fragile sites.

MAIN TYPES OF EXPLOSIVES

Black powder (gunpowder) = potassium nitrate, sulphur and carbon; slow expansion, used to extract dimension stone (section 39).
Dynamite = 20–60% nitroglycerine, with ammonium nitrate (or nitrocellulose in gelatin dynamites); greater power, lift, fracturing and vibration.
ANFO = 94% ammonium nitrate and 6% fuel oil; cheap, safe to handle, dissolves in water; 40% less powerful than dynamite; more efficient in weak rock

BLAST VIBRATIONS

Normally measured as peak particle velocity (ppv) in the 5–20 Hz range.
General safe limit of 50 mm/s ppv may be modified for different structures and relates to charge weight (of dynamite) and distance to the structure; graph refers to typical conditions, which may vary slightly with the local geology.
Reduce vibration levels with delayed firing (< 10 ms between smaller individual charges within the round) or low density explosives.

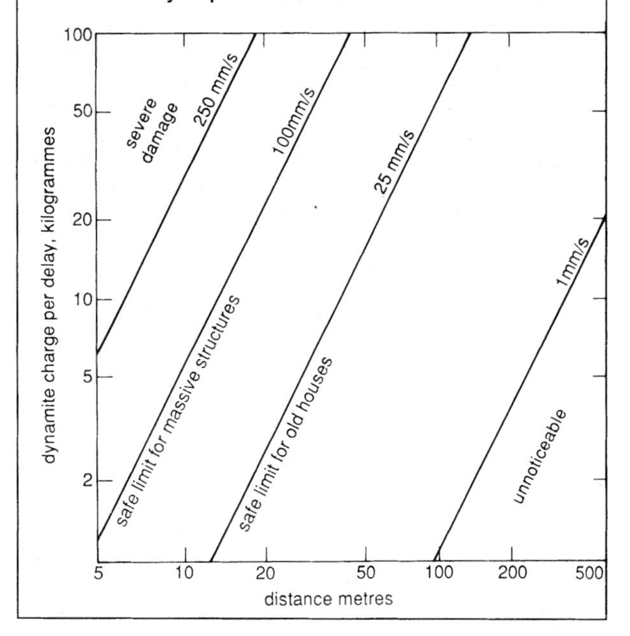

GROUNDWATER CONTROL

Good drainage of any site excavation is essential as it is normally the most economical means of rock slope stabilization (section 36).
Pumped drainage allows excavation below water table.

Horizontal drains, with slight gradient to provide gravity flow; bored holes 100 mm diameter, 10 m spacing, 50 m long, drain off site or into pumped sump.

Vertical well points, for temporary dewatering while site is worked inside coalesced cones of depression; well points are jetted into soils, 1–2 m spacing, can lift 5 m with surface vacuum pump. Submersible pumps in bored holes in rock or soil can be deeper – capacity and spacing depends on ground permeability and flows. Seepage flowing away from excavation improves slope stability.

Groundwater barriers permit dry excavation without lowering surrounding water table; sheet piles, concrete diaphragm walls, grouted zones or ground freezing, in order of rising cost; grouting or freezing can also control rising groundwater in thick aquifers.

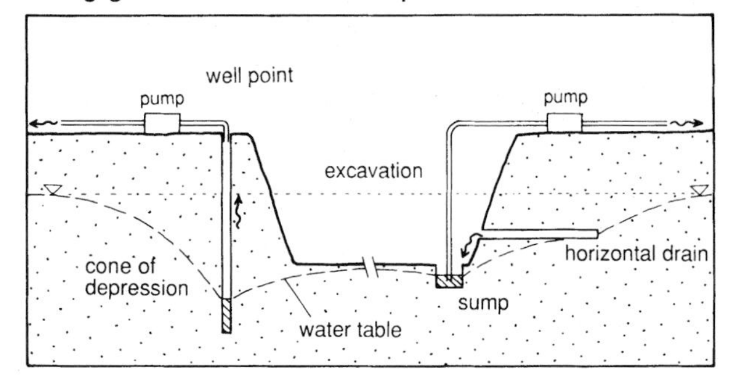

39 Tunnels in Rock

TUNNEL EXCAVATION

Choice of three methods. All cost £1–20M/km.
Drill and blast: in any rock, and for all large caverns.
Roadheader: machine mounted with a rotary milling head (specific to rock type); flexible system best for use in rock < 60 MPa UCS.
Tunnel boring machine: increasingly used, though uneconomic in tunnels < 1 km long.
Shaft sinking involves tedious debris clearance; raise boring is self clearing and often easier.

DRILL AND BLAST TUNNELLING

Full face, benched or crown heading and breakout where bad rock needs support.
Jumbo machines have 2–5 drills on arms; can swing to drill bolt holes.
Drill holes converge round a central wedge which is blasted out first; delay fuses (1–100 ms) then break outer zone into the central hole.
Smooth blasting leaves clean walls with perimeter holes spaced < 1 m, lightly charged and fired simultaneously.

TATES' CAIRN TUNNEL, HONG KONG, 1989
Progress in tunnel 10·7 m wide, 8 m high, with full face working in strong granite.
Two 10 h shifts/day, advance 60 m/week.
Drilling (3 h): round of 90 holes, 50 mm diam., 4·5 m deep, each hole takes 3 min with jumbo.
Charging and firing (2·5 h): 4·5 kg dynamite per hole.
Mucking (4 h): front loader fills 20 t dumptruck in 2 min; 1000 t per round, bulks 50%.

TUNNEL BORING MACHINES (TBM)

Full face rotating head, up to 9 m diameter, armed with roller discs or chisel picks, turning 2–10 rpm.
Now capable of working through hard rock.
Progress by jacking against side gripper pads or the installed concrete segment lining.
Advance: 30 m/day in soft ground, less in hard rock.
Cannot vary diameter; tightest curve = 300 m radius.
Some work as earth pressure balance shields, with a bulkhead providing face support in soft ground.

DIFFICULT GROUND CONDITIONS

Faults cause many problems – broken ground, increased water flow, and maybe change of rock type.
Groundwater is difficult in high flows; worst are karst fissures with up to 500 l/s, may need bulkhead sealing and/or flow diversion.
Overbreak in hard fractured rocks is worst in sedimentary and metamorphic along vertical strike.
Squeezing ground = plastic flow, mostly in clays and shales where UCS/overburden stress < 2.
Rockbursts occur mainly at depths > 600 m in rock with UCS > 140 MPa.
Swelling ground = wall closure due to any increased water content in clays.
Temperature of ground increases 2–4°C/100 m depth.
Rockhead is major hazard zone; underwater tunnels keep > 20 m rock cover.
Stress reduction in roof causes loosening of jointed rock and potential delayed failure.
Vertical stress in tunnel walls rises to 3 times overburden load.

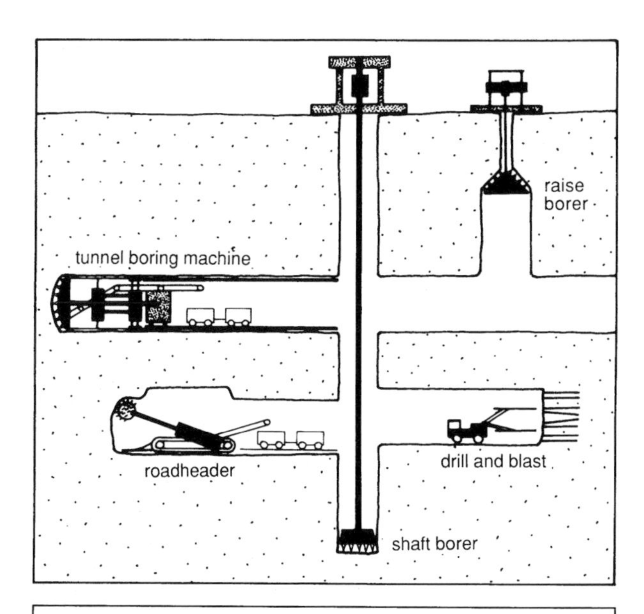

SEIKAN TUNNEL, JAPAN, 1985
Rail tunnel 54 km long, reaches 250 m below sea level.
In difficult mixture of faulted volcanic rocks, with UCS varying 3–150 MPa.
Cut by drill and blast, with 200 mm shotcrete lining, and steel arches in soft rock; grouting in fault zones after major flooding; cost £2500M.
Kept 100 m below seabed; advance probes all the way.

CHANNEL TUNNEL, EUROPE, 1992
Rail tunnel 50 km long, twin bores 7·6 m diameter.
In impermeable chalk marl with low fissure density, UCS = 5–9 MPa, close to ideal tunnelling medium.
Cut by 8·7 m TBM, and lined with precast concrete segments 360–540 mm thick; cost £900M.
Kept 20 m sound rock between crown and seabed.

ADVANCE GROUND IMPROVEMENT

Spiling strengthens arch of rock with advance fans of rockbolts 10° from tunnel axis.
Grout may be injected through similar fan of holes ahead of tunnel or into exposed zones of weakness.
Drainage control from surface, by well pointing, grouting or freezing, usually only in soils at shallow depths.

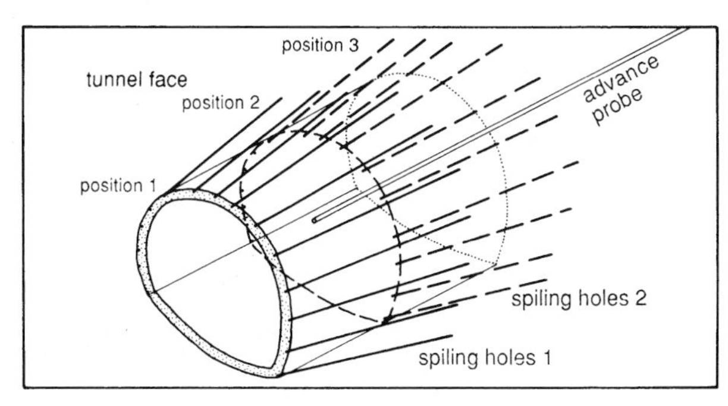

GROUND INVESTIGATION

Costs 0·5–3% of project, most in mixed weak rock.
Variable ground may warrant advance probes, 20–30 m ahead of face.
Seismic geophysics (section 22) can locate zones of poor rock by reduced velocity, beneath land or water.

Rock Tunnel Support Systems

PASSIVE SUPPORT

Use steel ribs (colliery arches) cast with concrete or precast concrete segments (often placed by TBM) grouted on outside.
Rock may impose high stress on parts of support.
Cast concrete, on travelling formwork, now mainly used as secondary linings for road tunnels.

ACTIVE SUPPORT

Strengthens the rock, mainly with rockbolts and shotcrete, to create a stable self-supporting arch within the rock mass over a tunnel opening.
Better than passive support in all but weak soils.
Norwegian Tunnelling Method successfully defines the required support; it is an adaptable system (based on experience) which can be applied to variable rock conditions. Extent of bolting, with or without shotcrete, is defined by rock mass quality and tunnel width.
New Austrian Tunnelling Method (NATM) allows limited rock deformation around tunnel; this redistributes stress to achieve new stable state, but is not enough to permit loosening and weakening. Bolts and thin flexible shotcrete lining are rapidly installed to take only a part of the load; deformation (of 10–100 mm) is permitted as rock takes up remaining stress, before secondary lining is installed.

SHOTCRETE

Concrete 20–200 mm thick sprayed onto rock wall.
Ideal as it interacts with rock, turning fractured rock into stronger rock mass in surface zone of peak stress.
Placed at 10 m^3/h, gives rapid flexible support in NATM, and protects rock from exposure weakening.
Layer 150 mm thick on 10 m diameter tunnel, as lining in compression, safely carries 450 kN load (= 20 m failed rock overburden).
Reinforced for tensile strength with steel weldmesh.

Fibrecrete contains 50–80 kg/m^3 of steel fibres each 40–50 mm long; better reinforcement than steel mesh. Reinforced shotcrete is tied to raised flanges on rockbolts to create integral rock–shotcrete arch.

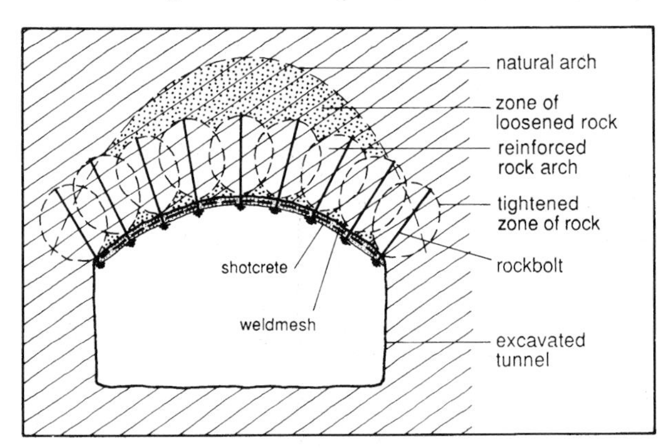

ROCKBOLTS

Mostly 2–5 m long, in 35 mm holes; load to 100 kN.
Three main types of fixing:

- Expansion shell, takes immediate load; cheapest.
- Grouted in with resin or cement; strongest.
- Friction, various types; Swellex deformed steel tube is expanded in hole by 30 MPa water pressure, takes immediate load; simplest.

Roof bolting should have:

- length = 1·4 + (tunnel width/5) m;
- spacing = length/2 and < (3 x joint spacing);
- tension against cap plate to 60% capacity.

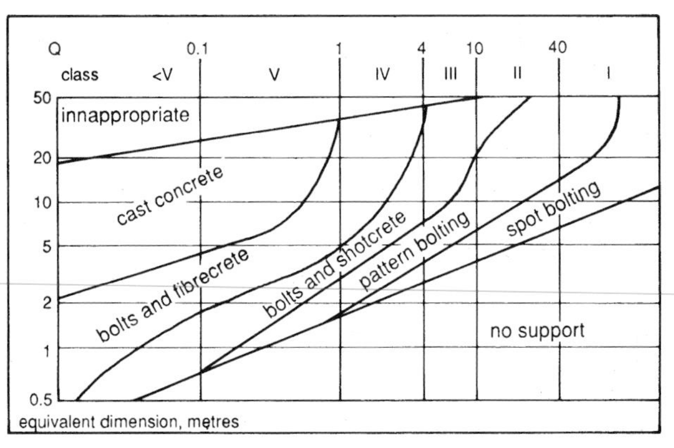

SELECTION OF SUPPORT SYSTEM

Chart above shows appropriate support system based on tunnel dimension and rock mass quality.
Q value defines rock mass properties on Norwegian system, summarized in section 40.
Class refers to rock mass (section 25).
Equivalent dimension = actual roof span/ESR.
Excavation Support Ratio reflects needs of safety:

Temporary mine openings	ESR = 3·0
Water tunnels, pilot tunnels	1·6
Access and minor rail tunnels	1·3
Major road and rail tunnels	1·0
Underground stations as public areas	0·8

Within each support system, dimensions vary broadly as in graphs below; further detail is related to rock class and tunnel size within full Norwegian classification.

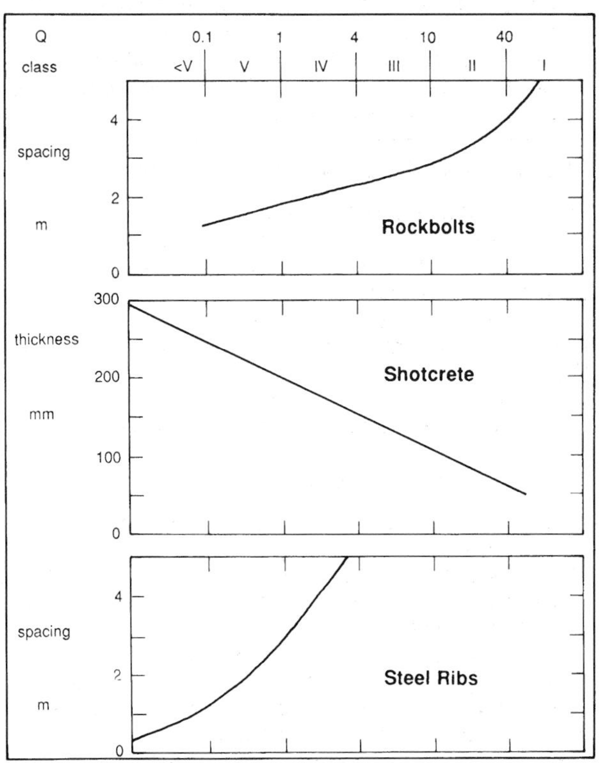

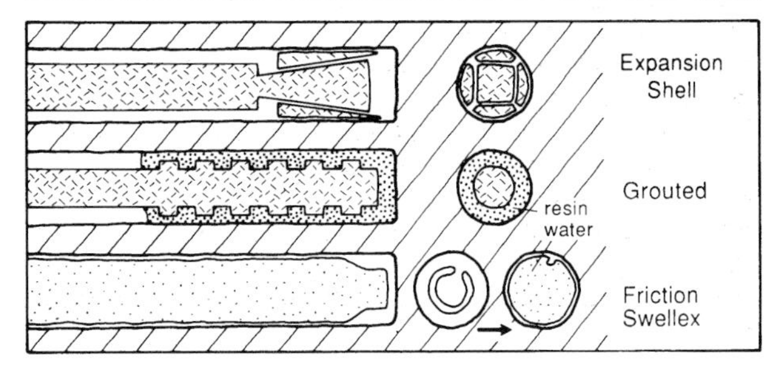

40 Stone and Aggregate

Demand for stone is around 200 Mt/y in Britain.
Less than 1 Mt/y is used as dimension stone.
All the rest is aggregate of which roughly

- Half goes to roads, half to concrete construction;
- Over half is crushed rock, rest is sand and gravel.

AGGREGATE PROPERTIES

Rock strength is the prime requirement – usually needs UCS > 100 MPa or 10% FV > 100 kPa.
Main demand is for particle sizes 5–50 mm; screened at quarry and normally sold as single size.
Quality is specified by aggregate tests (below).

Aggregate trade groups (opposite)
Provide useful classification as they group materials by properties and not by geological origin, and therefore identify some general features:

- Basalt and gabbro show good tar bonding to the iron in their minerals.
- Gritstone and granite have very high PSV and their rough surfaces adhere well to their binders.
- Flint and porphyry have smooth surfaces which can cause poor bonding, may have sharp edges and are more prone to alkali reaction.

Some younger and softer limestones and sandstones fall outside the grouping, along with all other weak sedimentary rocks which cannot be used as aggregate.

Lightweight aggregate may be provided by young volcanic pumice or some porous synthetics including pulverized fuel ash.

Concrete aggregate needs 10% FV > 100 for structural work, though 10% FV > 60 is adequate for some ground concrete; alkali reaction potential is important; rounded aggregate makes a concrete which flows more easily.

Road wearing course needs 10% FV > 100, good tar bonding and PSV > 60, though PSV also depends on the sorting and the binder, and different road types demand different values.

Railway ballast needs 10% FV > 100 and AIV < 18.

TYPICAL AGGREGATE PROPERTIES

Material	Location	10% FV	AIV	AAV	PSV
Granite	Dartmoor	280	16	5	60
Dolerite	N. Pennines	360	10	4	60
Greywacke	Pennines	220	14	7	65
Gritstone	Peak District	90	40	26	74
Limestone	Pennines	120	20	12	40
Flint	Thames gravel	450	23	1	35

AGGREGATE IMPURITIES

These must generally be avoided, and may be limited by contract specifications.

- Clay and mica: weak, absorptive, expansive.
- Opaline silica: alkali reaction.
- Pyrite: weathers to sulphuric acid and rust.
- Coal and lignite: react with bitumen binders.
- Organic (shell and plant): weak and reactive.
- Salt: corrosion, efflorescence, expansion.
- Sulphate: expansion, efflorescence.

ALKALI AGGREGATE REACTION

A mechanism of concrete deterioration due to reaction between certain types of aggregates and alkaline pore fluids in the concrete. A silica gel is formed which absorbs water and expands, thereby cracking the concrete. This may take 5–10 years to develop.
Resultant cracks allow more water and salt in which cause even more corrosion of reinforcing steel and deterioration of concrete.
Main reaction is with hydrated, opaline silica in aggregate – mostly in young acid volcanics, tuffs and some cherts. Reaction can also be with mixture of calcite, dolomite and illite in some limestones.
Avoid by using aggregates already successfully used, or keep alkalies <0·6% if acid volcanics or dolomitic limestone have to be used.

AGGREGATE TESTS

Various tests are used for contract specifications.
Summarized in table, with guidelines to indicate good and bad values, and limits generally used for wearing course roadstone.
Most test procedures are defined in British Standard 812; other countries have their own similar tests; American tests are in volumes of the annual book of ASTM Standards.
All tests are carried out on prepared aggregate samples.
Strength is indicated most closely by ACV. This value is rarely quoted directly; instead, 10% fines value is graphically determined from a series of ACV tests with different loads.
10% Fines Value is in kPa and is numerically just a little higher than UCS in MPa.
CBR is field test; has been extended to testing sub-grade soils, which have lower values than aggregate. Plastic clays generally have CBR < 10; sandy soils typically have CBR = 10–40.
Particle angularity, surface roughness and thermal expansion may also be measured and defined in some situations.

STANDARD AGGREGATE TESTS

Aggregate property	Test procedure (detailed in the appropriate part of BS 812)	Range of values		Road stone
		Good	Poor	
Aggregate impact value (AIV)	% fines lost by hammering on standard rig	5	35	<20
Aggregate abrasion value (AAV)	% loss by abrasion on standard test	1	25	<10
Polishing stone value (PSV)	Frictional drag recorded on pendulum swing	70	30	>60
Aggregate crushing value (ACV)	% fines lost by uniform load crushing on standard rig	5	35	
10% fines value (10% FV)	Load on standard ACV test rig to give 10% fines loss	400	20	>100
Flakiness index	Weight % particles with minimum thickness < 60% mean	20	70	<3
Water absorption	Weight % increase after immersion in water for 24 hours	0·2	10	<2
Frost heave	Heave of air-cooled column of sample standing in water	3	20	<1
California bearing ratio (CBR)	Resistance to plunger penetration, compared to standard	100	60	>90

CRUSHED ROCK

Angular rock chippings are better for roadstone.
Quality control is simple where a single rock mass is being quarried.
Weak impurities are reduced to fines in crushing process so are easily removed by screening.
Main costs are for blasting and crushing.
Selection often based on distance from quarry to site; land transport costs soon exceed quarry costs.
Remote coastal quarries overcome environmental and cost problems; Glensanda granite quarry in NW Scotland can economically ship aggregate to Bristol and London, and to Texas coast.

NATURAL AGGREGATES

Alluvial gravels are the most important resource, from floodplains and terraces.
Dredged marine gravels and some glaciofluvial gravels are also used; glacial till is too poorly sorted.
Rounded gravel particles are better for concrete.
Quality control is more difficult; alluvial gravel consists of any rocks within the river catchment, so may be varied; River Trent gravels are mainly strong quartzite from Triassic conglomerates, but also contain coal fragments; coal is removed by washing (density separation in turbulent water).
Main costs are for overburden stripping and screening.

AGGREGATE TRADE GROUP CLASSIFICATION

Group	Including	Characteristics	Quality
Basalt	dolerite	strong, fine-grained, basic igneous	good
Gabbro		strong, coarse-grained, basic igneous	
Porphyry	rhyolite	strong, fine-grained, acid igneous	
Granite	gneiss	strong, coarse-grained, acid igneous	good
Hornfels		strong, fine-grained, uncleaved metamorphic	good
Schist	slate	flakey, sheared or cleaved metamorphic	poor
Quartzite		strong, metamorphized sandstone	rare
Limestone	marble	the stronger limestones and dolomites	good
Gritstone	greywacke	the stronger, well-cemented sandstones	good
Flint	chert	fine-grained silica, mostly as gravel	
Artificial		any synthetic slags	
Reject		all the soft sedimentary rocks	useless

Slate quarry in Snowdonia, Wales

Dimension Stone

This is stone used in large, uncrushed blocks; rock must have low fracture density to permit extraction of large blocks.

Construction stone may be any locally available rock with UCS > 50 MPa. Mostly limestones, sandstones and granites, now largely replaced by concrete.

Freestone is best used for carved ornamental work as it has no preferred fracture direction nor any planar weaknesses.

Cladding stone is used in 10–20 mm thick sheets as facing on concrete. Needs UCS > 100 MPa, should be attractive and must often take a polish. Dominated by marbles and granites.

Armour stone is large, uncut, chunk rock used for erosion defence. Exposed marine sites need UCS > 150 MPa in large blocks, so granite used; but density > 2·8 t/m^3 (in dolerite and gneiss) may be specified. Limestone with UCS > 100 MPa acceptable for smaller waves on lakes, and for protective riprap of smaller sized blocks on earth dam faces.

Sandstone: Much used in the past, less today. Cannot polish and rough surface blackens with soot. York Stone is any Carboniferous sandstone from the Pennines of northern England.

Flagstone: Variety which splits into bedding slabs 40–50 mm thick. Good for paving flags; also used on roofs but very heavy due to thickness.

Granites: Very strong, suitable for all uses, but hard and therefore expensive to polish.

Larvikite: Variety with internal reflections from feldspar crystals; makes attractive dark cladding.

Limestones: Strong limestones which take a polish are known in the trade as marble; they are softer and cheaper to polish than granites.

Marble: Along with the strong limestones make excellent cladding and architectural stone.

Softer limestones: Do not polish, but freestones make excellent building stone – notably Cotswold and Portland Stones of southern England. Some may weather badly, but these case-harden due to redeposition by porewater on exposure.

Travertine: Soft, easily carved, best for internal use.

Slate: Strong and very durable; may be split along cleavage, using hammer and chisel, into sheets 4–6 mm thick; ideal for roofing. Best Welsh slate has highest flexural strength and can be cleaved to 1 mm thick. Some used cut for architectural work.

High value of best dimension stone means it can be shipped greater distances and is widely available. Notable examples are Carrara marble from Italy, larvikite from Larvik in Norway, slate from Wales, Rock of Ages granite from Vermont in USA.

Appendices

DESCRIPTION OF ROCK MASS QUALITY BY THE NORWEGIAN Q SYSTEM

Ratings are determined visually for the six parameters. The Q value is then calculated as:

$$Q = (RQD/J_n) \times (J_r/J_a) \times (J_w/SRF)$$

Q values are compared to rock mass properties and other ratings systems in section 25.

These tables only summarize the Q parameters. The full system, with many refinements, and its application to the design of tunnel support systems is in Barton, N., Lien, R. and Lunde, J. (1974) Engineering classification of rock masses for tunnel design. *Rock Mechanics*, **6**(4), 189–236.

Rock Quality Designation	**RQD**
Values from borehole data	10–100
If RQD <10, use 10 to calculate Q	10

Joint Set Number	J_n
Massive, no or few joints	0·5–1·0
One joint set; if random also, add 1	2
Two joint sets; if random also, add 2	4
Three joint sets; if random also, add 3	9
Four or more joint sets	15
Crushed rock, earthlike	20

Joint Roughness Number	J_r
Discontinuous joints	4
Rough or irregular, undulating	3
Smooth and undulating	2
Rough and planar	1·5
Smooth and planar	1·0
Slickensided and planar	0·5
No rock wall contact across gouge	1·0

Joint Alteration Number	ϕ_r	J_a
Rock walls in contact:		
Clean, tight joints	> 25°	0·75–1·0
Slightly altered joint walls	25 – 30°	2
Silty or sandy clay coatings	20 – 25°	3
Soft clay coatings	8 – 16°	4
Gouge or filling < 5 mm thick		
Sandy particles or fault breccia	25 – 30°	4
Stiff clay gouge	16 – 24°	6
Soft or swelling clay gouge	6 – 12°	8–12
Thick continuous clay zones	6 – 24°	10–20

Joint Water Factor	Water pressure	J_w
Dry or minor flow	< 100 kPa	1·0
Medium inflow	100 – 250 kPa	0·66
Large flow in sound rock	250 – 1000 kPa	0·5
Large flow washing out joint infills	250 – 1000 kPa	0·33
Very high flows	> 1000 kPa	0·2–0·05

Stress Reduction Factor		**SRF**
Fractured rock prone to loosening		
Multiple weakness zones with clay, loose rock		10
Multiple weakness zones, no clay, loose rock		7·5
Single weakness zones, cover depth > 50 m		5
Single weakness zones, cover depth < 50 m		2·5
Loose open joints		5
Sound rock	UCS/major stress	
Low stress, near surface	> 200	2·5
Medium stress	200 – 10	1·0
High stress, tight structure	10 – 5	0·5 – 2·0
Mild–heavy rock bursts	< 5	5 – 20
Mild squeezing or swelling rock		5 – 10
Heavy squeezing or swelling rock		5 – 20

Cover photograph. The western portal of the Penmaenbach road tunnel, completed in 1989 for the improvements to the North Wales coast road. The tunnel was cut through 660 m of class II and class I rock, a strong rhyolite with three sets of widely spaced joints. Removal of scree and weathered rock from the site of the portal allowed stress relief within the rhyolite. This was countered by 48 rock anchors each 7 m long, installed before cutting the tunnel, to prevent opening of the joints. Extension of the secondary concrete lining beyond the face, and the masonry rock catches above, are to prevent small stonefall reaching the road.

ABBREVIATIONS AND NOTATION

ABP	acceptable bearing pressure
C	centigrade
c	cohesion
c′	effective cohesion
c_r	residual cohesion
CPT	cone penetration test
E	Young's modulus of elasticity
Fe	iron
g	acceleration due to gravity
GPa	1 000 000 000 Pascals
GPR	ground probing radar
ha	hectare, area 100 square metres
Hz	Hertz, frequency of 1 cycle per second
K	coefficient of permeability
kg	kilogramme
km	kilometre
kN	1000 Newtons
kPa	1000 Pascals
LI	liquidity index
LL	liquid limit
l/s	litres per second
M	1 000 000
M	earthquake magnitude
m	metre
mm	millimetre
MN	1 000 000 Newtons, about 100 tonnes
MPa	1 000 000 Pascals, about 100 t/m^2
m_v	coefficient of compressibility
My	million years
N	Newton, force to accelerate 1 kg to 1 m/s^2, so that, under influence of gravity, 1 kg = 9.81 N
NATM	new Austrian tunnelling method
NGR	national grid reference
OD	ordnance datum, effectively mean sea level
Pa	Pascal, unit of stress or pressure, N/m^2
PFa	pulverized fuel ash
PI	plasticity index
PL	plastic limit
PLS	point load strength
Q	rating value of rock mass, Norwegian system
Q	flow
RMR	rock mass rating on Geomechanics system
RQD	rock quality designation
s	second
SPB	safe bearing pressure
SiO_2	silica, silicon dioxide
SPT	standard penetration test
t	tonne
TBM	tunnel boring machine
u	pore water pressure
UCS	unconfined compressive strength
U100	100 mm diameter soil sample
w	water content
3-D	three-dimensional
γ	unit weight of material
μ	micro, a millionth part
Σ	sum of
σ	stress
σ'	effective stress
σ_n	normal stress
τ	shear strength
ϕ	internal friction angle
ϕ'	effective internal friction angle
ϕ_r	residual internal friction angle

FURTHER READING

As references are not cited throughout the text, the most relevant literature is cited below for each section of this book. The first citations are the major source books and useful papers in the subject area. These are followed by the primary papers on the case studies, cited in the sequence in which they are mentioned within the text.

01

Kiersch, G. A. (ed), 1991. The heritage of engineering geology: the first hundred years. *Geol. Soc. Am. Centennial Special*, Volume 3, 605pp.

Duff, D. (ed), 1993. *Holmes' principles of physical geology*. Nelson Thornes, 816pp.

Legget, R. F. & Hatheway, A. W., 1988. *Geology and engineering*. McGraw Hill, 613pp.

Bell, F. G., Cripps, J. C. & Culshaw, M. G., 1995. The significance of engineering geology to construction. *Geol. Soc. Eng. Geol. Spec. Publ.*, **10**, 3–29.

Bennett, M. R. & Doyle, P., 1997. *Environmental geology*. Wiley, 501pp.

Murck, B. W., Skinner, B. J. & Porter, S. C., 1996. *Environmental geology*. Wiley, 534pp.

Waltham, T., 1978. *Catastrophe: the violent Earth*. Macmillan, 170pp.

Google.co.uk – the best search engine for geological data.

02

Vernon, R., 2000. *Beneath our feet: the rocks of planet Earth*. Cambridge University Press, 216pp.

Goodman, R. E., 1993. *Engineering geology: rock in engineering construction*. Wiley, 412pp.

Thorpe, R. & Brown, G., 1985. *The field description of igneous rocks*. Wiley, 154pp.

03–04

Greensmith, J. T., 1989. *Petrology of the sedimentary rocks*. Unwin Hyman, 262pp.

Blatt, H., 1992. *Sedimentary petrology*. Freeman, 514pp.

Goodman, 1993. *see #02*.

Vernon, 2000. *see #02*.

Tucker, M. E., 1996. *Sedimentary rocks in the field*. Wiley, 153pp.

05

Mason, R., 1990. *Petrology of the metamorphic rocks*. Unwin Hyman, 230pp.

Goodman, 1993. *see #02*.

Vernon, 2000. *see #02*.

Fry, N., 1984. *The field description of metamorphic rocks*. Wiley, 110pp.

06

Park, R. G., 1997. *Foundations of structural geology*. Nelson Thornes, 202pp.

Twiss, R. J. & Moores, E. M., 1992. *Structural geology*. W H Freeman, 532pp.

Spencer, E. W., 1988. *Introduction to the structure of the Earth*. McGraw-Hill, 551pp.

Ramsay, J. G. & Huber, M. I., 1987. *The techniques of modern structural geology; volume 2: folds and fractures*. Academic Press, 400pp.

07–08

Lisle, R. J., 1995. *Geological structures and maps*. Butterworth Heinemann, 104pp.

Thomas, P. R., 1991. *Geological maps and sections for civil engineers*. Blackie, 106pp.

McClay, K., 1987. *The mapping of geological structures*. Wiley, 161pp.

Dearman, W. R., 1991. *Engineering geological mapping*. Butterworth Heinemann, 580pp.

09

Kearey, P. & Vine, F. J., 1996. *Global tectonics*. Blackwell, 333pp.

Ollier, C. & Pain, C., 2000. *The origins of mountains*. Routledge, 345pp.

10

Bolt, B. A., 1999. *Earthquakes.* W H Freeman, 366pp.

US Geological Survey, 2001+. *Earthquakes hazards program.* www.usgs.gov/earthquake

Francis, P., 1993. *Volcanoes.* Clarendon, 443pp.

Smithsonian Institution, 2001+. *Global volcanism program.* www.volcano.si.edu/gvp

11

Duff, P. M. D. & Smith, A. J. (eds), 1992. *Geology of England and Wales.* Geological Society, 651pp.

12

Stearn, C. W. (ed), 1979. *Geological evolution of North America.* Wiley, 540pp.

13

Bland, W. & Rolls, D., 1998. *Weathering: an introduction to the scientific principles.* Arnold, 271pp.

Gerrard, A. J., 1988. *Rocks and landforms.* Unwin Hyman, 319pp.

14

Ahnert, F., 1996. *Introduction to geomorphology.* Arnold, 352pp.

Knighton, D., 1998. *Fluvial forms and processes: a new perspective.* Arnold, 383pp.

Forster, A., Culshaw, M. G., Cripps, J. C., Little, J. A. & Moon, C. F., 1991. Quaternary engineering geology. *Geol. Soc. Eng. Geol. Spec. Publ.*, **7**, 724pp.

Brookes, A., 1985. River channelization: traditional engineering, physical consequences and alternative practices. *Prog. Phys. Geog.*, **9**, 44–73.

Kidson, C., 1953. The Exmoor storm and the Lynmouth floods. *Geography*, **38**, 1–9.

15

Menzies, J. (ed), 1996. *Postglacial environments: sediments, forms and techniques.* Butterworth Heinemann, 598pp.

Bennett, M. R. & Glasser, N. F., 1996. *Glacial geology and ice sheet landforms.* Wiley, 376pp.

Eyles, N. & Sladen, J. A., 1981. Stratigraphy and geotechnical properties of weathered lodgement tills in Northumberland. *Quart. Journ. Eng. Geol.* **14**, 129–141.

Bryan, A., 1951. *Accident at Knockshinnoch Castle Colliery, Ayrshire: report.* HMSO, London, 48pp.

16

Livingstone, I. & Warren, A., 1996. *Aeolian geomorphology: an introduction.* Longman, 211pp.

Cooke, R. U., Brunsden, D., Doornkamp, J. C. & Jones, D. K. C., 1982. *Urban geomorphology in drylands.* Oxford University Press, 370pp.

Watson, A., 1985. The control of wind blown sand and moving dunes. *Quart. Journ. Eng. Geol.*, **18**, 237–252.

Ballantyre, C. K. & Harris, C., 1994. *The periglaciation of Great Britain.* Cambridge University Press, 330pp.

Coxon, R. E., 1986. *Failure of Carsington embankment.* Dept. Environment Report, HMSO, London, 180pp.

Skempton, A. W. & Vaughan, P. R., 1993. The failure of Carsington dam. *Geotechnique*, 43, 151–173.

Waltham, T. & Fookes, P., 2001. Ice wedges of the Dalton Highway, Alaska. *Q. Journ. Eng Geol. Hydro.*, **34**, 65–70.

17

Pethick, J., 1984. *An introduction to coastal geomorphology.* Arnold, 260pp.

Haslett, S. K., 2000. *Coastal systems.* Routledge, 218pp.

Barrett, M. G. (ed), 1992. *Coastal zone planning and management.* Thomas Telford, 327pp.

British Standards, 1991. Maritime structures: guide to the design and construction of breakwaters. ***BS*** 6349, 92pp.

18

Grassington, R., 1998. *Field hydrogeology.* Wiley, 248pp.

Price, M., 1996. *Introducing groundwater.* Nelson Thornes, 304pp.

Mackay, R., Riley, M. & Williams, G. M, 2001. Simulating groundwater contaminant migration at Villa Farms lagoons. *Q. Journ. Eng. Geol. Hydrogeol.*, **34**, 215–224.

19

British Standards, 1999. Code of practice for site investigations. ***BS*** 5930, 204pp.

Simons, N., Menzies, B. & Matthews, M., 2001. *A short course in geotechnical site investigation.* Thomas Telford, 250pp.

Weltman, A. J. & Head, J. M., 1983. *Site investigation manual.* CIRIA, London, 144pp.

22

Milsom, J., 2000. *Field geophysics.* Wiley, 187pp.

Telford, W. M., Geldart, L. P. & Sheriff, R. E., 1991. *Applied geophysics.* Cambridge Univ. Press, 790pp.

23

Waltham, A. C., Vandenven, G. & Ek, C. M., 1986. Site investigations on cavernous limestone for the Remouchamps Viaduct, Belgium. *Ground Engineering*, 19 (8), 16–18.

24

Goodman, R. E., 1989. *Introduction to rock mechanics.* Wiley, 563pp.

Wyllie, D. C., 1999. *Foundations on rock.* Spon, 401pp.

25

Hoek, E., 2000+. *Practical rock engineering.* www.rockeng.utoronto.ca

Barton, N., Lien, R. & Lunde, J., 1974. Engineering classification of rock masses for the design of tunnel support. *Rock Mechanics*, **6**, 189–236.

Bieniawski, Z. T., 1974. Geomechanics classification of rock masses and its application in tunnelling. *Proc. 3rd Int. Cong. Rock Mech.*, **2** (2), 27–32.

Bieniawski, Z. T., 1989. *Engineering rock mass classifications.* Wiley, 251pp.

Hudson, J. A. & Harrison, J. P., 1997. *Engineering rock mechanics.* Pergamon, 444pp.

Wyllie, 1999. *see #24.*

British Standards, 1986. Code of practice for foundations. ***BS*** 8004, 160pp.

Bell, A. L., 1994. *Grouting in the ground.* Thomas Telford, 590pp.

26

Jefferson, I., Greenwood, J. & Frost, M., 2002. *Foundations of geotechnical engineering.* Spon, 96pp.

Craig, R. F., 1997. *Soil mechanics.* Spon, 486pp.

27

Waltham, A. C., 1989. *Ground subsidence.* Blackie. 202pp.

Ege, J. R., 1984. Mechanisms of surface subsidence resulting from solution extraction of salt. *Geol. Soc. Am. Reviews in Eng. Geol.*, **6**, 203–221.

Cooper, A. H. & Waltham, A. C., 1999. Subsidence caused by gypsum dissolution at Ripon, North Yorkshire. *Quart Journ. Eng. Geol.*, **32**, 305–310.

Stephens, J. C., Allen, L. H. & Chen, E., 1984. Organic soil subsidence. *Geol. Soc. Am. Reviews in Eng. Geol.* **6**, 107–122.

Hutchinson, J. N., 1980. The record of peat wastage in the East Anglia fenlands at Holme Post, 1846–1978 AD. *Journ. Ecology*, **68**, 229–249.

Samson, L. & La Rochelle, P., 1972. Design and performance of an expressway constructed over peat by preloading. *Can. Geotech. Journ.*, **9**, 447–466.

28

Peck, R. B. & Bryant, F. G., 1953. The bearing capacity failure of Transcona elevator. *Geotechnique*, 3, 201–214.

Burland, J., 1997. *Propping up Pisa.* Royal Academy of Engineering, London. 20pp.

Figueroa Vega, G. E., 1984. Land subsidence case history: Mexico. *Unesco Studies and Reports in Hydrology*, **40**, 217–232.

Zeevaert, L., 1957. Foundation design and behaviour of the Tower Latino Americana in Mexico City. *Geotechnique*, **7**, 115–133.

Carbognin, L. & Gatto, P., 1986. An overview of the subsidence of Venice. *Int. Ass. Hydrol. Sci. Publ.*, **151**, 321–328.

29

Sowers, G. F., 1975. Failures in limestone in humid subtropics. *Proc. Am. Sc. Civ. Eng.*, **101** (GT8), 771–787.

Waltham, 1989. *see #27.*

Culshaw, M. G. & Waltham, A. C., 1987. Natural and artificial cavities as ground engineering hazards. *Quart. Journ. Eng. Geol.*, **20**, 139–150.

Newton, J. G., 1987. Development of sinkholes resulting from man's activities in the eastern United States. *U. S. Geol. Surv. Circular*, 968, 54pp.

30

Littlejohn, G. S., 1979. Surface stability in areas underlain by old coal workings. *Ground Engineering*, **12**(2), 22–30.

Littlejohn, G. S., 1979. Consolidation of old coal workings. *Ground Engineering*, **12**(4), 15–21.

Waltham, 1989. *see #27.*

Culshaw & Waltham, 1987. *see #29.*

31

Whittaker, B. N. & Reddish, D. J., 1989. *Subsidence: occurrence, prediction and control.* Elsevier, 528pp.

National Coal Board, 1975. *Subsidence engineer's handbook.* NCB, 112pp.

32

Bromhead, E. N., 1997. *The stability of slopes.* Spon, 411pp.

Cruden, D. M. & Varnes, D. J., 1996. Landslide types and processes. 36–75 in Turner and Schuster, *see #32.*

Turner, A. K. & Schuster, R. L. (eds), 1996. Landslides: investigation and mitigation. *Transportation Research Board Special Report*, 247, 673pp.

Voight, B., 1978–9. *Rockslides and avalanches volumes 1 and 2.* Elsevier, 833+850pp.

Hadley, J. B., 1978. Madison Canyon rockslide, Montana, USA. 167–180 in Voight vol 1, *see #32.*

Voight, B., 1978. Lower Gros Ventre slide, Wyoming, USA. 113–166 in Voight vol 1, *see #32.*

Wieczorek, G. F., 1996. Landslide triggering mechanisms. 76–90 in Turner & Schuster, *see #35.*

Plafker, G. & Ericksen, G. E., 1978. Nevados Huascaran avalanches, Peru. 277–314 in Voight vol 1. *see #32.*

33

Bromhead, 1997. *see #32.*

Cruden, D. M. & Krahn, J., 1978. Frank rockslide, Alberta, Canada. 97–112 in Voight, *see #32.*

Ehlig, P. L., 1992. Evolution, mechanics and mitigation of the Portugese Bend landslide, Palos Verdes Peninsula, California. *Assoc. Eng. Geologists (Southern California Section), Spec. Publ.* **4**, 531–553.

Hendron, A. J. & Patton, F. D., 1986. A geotechnical analysis of the behaviour of the Vaiont slide. *Civil Eng. Practice* (Boston Soc. Civ. Eng.), **1**(2), 65–130.

34

Watson, R. A. & Wright, H. E., 1969. The Saidmarreh landslide, Iran. *Geol. Soc. Am. Spec. Paper*, **123**, 115–139.

Torrance, K. J., 1987. Quick clay. 447–473 in Anderson, M. G. & Richards, K. S. (eds). *Slope stability.* Wiley.

Crawford, C. B. & Eden, W. J., 1963. Nicolet landslide of November 1955, Quebec, Canada. *Geol. Soc. Am. Eng. Geol. Case Histories*, **4**, 45–50.

Anon., 1967. *Report of the tribunal appointed to enquire into the disaster at Aberfan on October 21, 1966.* HMSO, London, 148pp.

Siddle, H.J., Wright, M. D. & Hutchinson, J. N., 1996. Rapid failures of colliery spoil heaps in the South Wales coalfield. *Quart. Journ. Eng. Geol.*, **29**, 103–132.

35

Bromhead, 1997. *see #32.*

Norrish, N. I. & Wyllie, D. C., 1996. Rock slope stability analysis. 391–425 in Turner & Schuster, *see #32.*

Waltham, A. C. & Dixon, N., 2000. Movement of the Mam Tor landslide, Derbyshire, UK. *Quart. Journ. Eng. Geol. Hydrogeol.*, **33**, 105–123.

Ritchie, A. M., 1963. Evaluation of rockfall and its control. *Highway Research Board Record* (Canada), **17**, 13–28.

36

Bromhead, 1997. *see #32.*

Bromhead, E. N., 1997. The treatment of landslides. *Proc. Inst. Civ. Eng. Geotech. Eng.*, **125**, 85–96.

Hoek, E. & Bray, J., 1981. *Rock slope engineering.* Inst. Min. Met. London, 402pp.

Turner & Schuster, 1996. *see #32.*

Hutchinson, J. N., 1969. A reconsideration of the coastal landslides at Folkestone Warren, Kent. *Geotech-nique*, **19**, 6–38.

Wyllie, D. C. & Norrish, N. I., 1996. Stabilization of rock slopes. 474–504 in Turner & Schuster, *see #32.*

British Standards, 1994. Code of practice for earth retaining structures. ***BS*** 8002, 144pp.

British Standards, 2001. Geotextiles: characteristics required for earthworks, foundations and retaining structures. ***BS*** EN13251, 34pp.

British Standards, 1989. Code of practice for ground anchorages. ***BS*** 8081, 180pp.

37

Fookes, P. G., 1997. Geology for engineers: the geological model, prediction and performance. *Quart. Journ. Eng. Geol.*, **30**, 293–431.

Hoek, E., 1999. Putting numbers to geology – an engineer's approach. *Quart. Journ. Eng. Geol.*, **32**, 1–19.

Hutchinson, J. N., 2001. Reading the ground: morphology and geology in site appraisal. *Quart. Journ. Eng. Geol. Hydrogeol.*, **34**, 7–50.

Davison, L., Fookes, P., Baynes, F. & Hutchinson, J., 2001+. Total geological history: a model approach. www.uwe.ac.uk/geocal/totalgeology

Charles, J. A., 1993. *Building on fill: geotechnical aspects.* Building Research Establishment Report, 163pp.

Hester, R. E. & Harrison, R. M. (eds), 1997. *Contaminated land and its reclamation.* Thomas Telford, 160pp.

British Standards, 2001. Investigation of potentially contaminated land. ***BS*** 10175, 82pp.

38

Franklin, J. A. & Desseault, M. B., 1991. Rock engineering. McGraw Hill, 401pp.

Hudson & Harrison, 1997. *see #25.*

Peffifer, G. S. & Fookes, P. G., 1994. A revision of the graphical method for assessing the excavatability of rock. *Quart. Journ. Eng. Geol.*, **27**, 145–164.

Lutton, R. J., Banks, D. C. & Strohm, W. E., 1979. Slides in Gaillard Cut, Panama. 151–224 in Voight vol 2. *see #32.*

Bhandari, S., 1997. *Engineering rock blasting operations.* Balkema, 375pp.

39

Franklin, J. A. & Desseault, M. B., 1991. *Rock engineering applications.* McGraw Hill, 431pp.

Muir Wood, A., 2000. *Tunnelling: management by design.* Spon, 307pp.

Tsuji, H., Sawada, T. & Takizawa, M., 1996. Extraordinary inundation accidents in the Seikan undersea tunnel. *Proc. Inst. Civ. Eng. Geotech. Eng.*, **119**, 1–14.

Harris, C. S., Hart, M. B., Varley, P. M. & Warren, C. D., 1996. *Engineering geology of the Channel Tunnel.* Thomas Telford, 520pp.

Barton, N. & Grimstad, E., 1994. Rock mass conditions dictate choice between NMT and NATM. *Tunnels and Tunnelling*, October, 39–42.

British Standards, 1996. Specification for rock bolting in coal mines. ***BS*** 7861, 38pp.

40

McNally, G. H., 1998. *Soil and rock construction materials.* Spon, 403pp.

Latham, J. P. (ed), 1998. Advances in aggregates and armourstone evaluation. *Geol. Soc. Eng. Geol. Spec. Publ.*, **13**, 201pp.

British Standards, 1992. Specifications for aggregates from natural sources for concrete. ***BS*** 882, 12pp.

Jefferson, D. P., 1993. Building stone: the geological dimension. *Quart. Journ. Eng. Geol.*, **26**, 305–319.

British Standards, 1976. Code of practice for stone masonry. ***BS*** 5390, 44pp.

Index